Electrode
and
Corrosion
Physics

Electrode
and
Corrosion
Physics

Anthony T Paxton

Imperial College London, UK
King's College London, UK

World Scientific

NEW JERSEY · LONDON · SINGAPORE · BEIJING · SHANGHAI · HONG KONG · TAIPEI · CHENNAI · TOKYO

Published by

World Scientific Publishing Europe Ltd.

57 Shelton Street, Covent Garden, London WC2H 9HE

Head office: 5 Toh Tuck Link, Singapore 596224

USA office: 27 Warren Street, Suite 401-402, Hackensack, NJ 07601

Library of Congress Cataloging-in-Publication Data
Names: Paxton, A. T. (Anthony T.), author.
Title: Electrode and corrosion physics / Anthony T Paxton, Imperial College London, UK ;
  King's College London, UK.
Description: New Jersey : World Scientific, [2024] | Includes bibliographical references and index.
Identifiers: LCCN 2023053693 | ISBN 9781800615489 (hardcover) |
  ISBN 9781800615533 (paperback) | ISBN 9781800615496 (ebook for institutions) |
  ISBN 9781800615502 (ebook for individuals)
Subjects: LCSH: Electrochemistry. | Electrolytic corrosion.
Classification: LCC QD553 .P37 2024 | DDC 541/.37--dc23/eng/20240224
LC record available at https://lccn.loc.gov/2023053693

British Library Cataloguing-in-Publication Data
A catalogue record for this book is available from the British Library.

For any available supplementary material, please visit
https://www.worldscientific.com/worldscibooks/10.1142/Q0458#t=suppl

Desk Editors: Logeshwaran Arumugam/Ana Ovey/Shi Ying Koe

Typeset by Stallion Press
Email: enquiries@stallionpress.com

To Sarah Felicity Angel

Preface

I expect almost everyone has seen the rusting hulk of a ship, car or machinery in an abandoned waterway, scrap yard or farm yard. Most of us have cause to deprecate the wastage due to corrosion, as all metals, but gold, strive to return to their ores. This inexorable march is the province of equilibrium thermodynamics. Possibly fewer people are aware that underlying all corrosion is electricity.

I have written this slim volume for material scientists, engineers and physicists, but I am hopeful that some chemists, even electrochemists, will find something profitable within these pages. The standpoint is that my reader will have a background in physics (solid state, electrostatics, statistical mechanics and even some quantum mechanics) but know chemistry only at the level of secondary or high school.

This was my own situation when I came to find in my research that I needed a deeper understanding of the literature and the textbooks in classical electrochemistry, and to some extent, this book is a travelogue of my journey towards that goal. In that sense, it is something of a personal account and so does not follow the usual progress of a more conventional textbook. Having said that, it is notable that unlike subjects such as statistical and quantum mechanics, electrostatics and thermodynamics, electrochemistry textbooks tend to be most individualistic and there does not appear to be a standard, agreed sequence in which to teach this subject.

It may have been more modern of me to write about fuel cells and batteries since these are the urgent technologies that require our immediate attention in the climate crisis. My intention is that the reader wishing to enter this field having the background indicated here will find the first twelve chapters of this book valuable in establishing the principles in their minds. And they may decide to proceed to the end having journeyed thus far and will be able to put the lessons in corrosion to good use, especially the study of chemical kinetics, in the design of efficient, cost-effective and sustainable devices.

To my mind, physics can only take us as far as uniform corrosion, which may disappoint the engineers in my readership. But again, my intention is that this text will instruct such readers in the fundamentals and in particular in the use of the Evans diagram which is the key tool to the description of crevice corrosion, galvanic corrosion, differential aeration and so on. Indeed pitting corrosion lies at the root of the most devastating and costly failures, but again this is a subject best tackled by the advanced and specialised literature.

The thermodynamics of charged entities is fraught with difficulties and pitfalls and I have attempted to bring out what is, and what is not, *measurable*. If it weren't presumptive of me, I would aver that some textbooks have fallen into confusion, not to say error, in this and also in the subject of interface thermodynamics. I only assume the reader has met the first and second laws in which the only work done is that against an external pressure (the pV term) and I have appended a tutorial in thermodynamics, adapted and extended from my lecture notes from a course on steel metallurgy at Imperial College some years ago.

I have included some problems but not in any systematic way. Standard texts in electrochemistry and physical chemistry contain many exercises, frequently taken from final year examination papers. I have decided to include a small number of worked problems which are intended to extend some of the topics covered. The problems are numbered by chapter, but some chapters have more than one problem and some have none.

The reader will notice that I have been much influenced by the lengthy, not to say prolix, volume of Bockris and Reddy, and I have also gained much from the texts of Fawcett and of Schmickler and Santos. A small Further Reading section can be found in Appendix D.

I would like to express my debt of gratitude to Mike Finnis, to Sasha Lozovoi and to Andrew Horsfield for their deep insights and to my editors, Arya Thampi, Ana Ovey, Koe Shi Ying and Logesh Arumugam, for their guidance and assistance in bringing this work into production.

London
September 2023

About the Author

Tony Paxton is a theoretical metallurgist and solid-state physicist. He started life as a laboratory technician, earning Distinction in City and Guilds welding craft practice. He obtained First Class Honours BMet and the Mappin Medal in Metallurgy at the University of Sheffield in 1984 and DPhil at the University of Oxford in 1987, where he became a Graduate Scholar and Admiral of Punts at Wolfson College. He worked at the Max-Planck-Institut für Festkörperforschung in Stuttgart and SRI International, Menlo Park California, before moving back to the Department of Metallurgy and the Science of Materials Oxford as SERC Advanced Fellow and Research Fellow Wolfson College in 1993. He took up a lectureship in Physics at Queen's University Belfast, becoming Professor of Theory and Modelling in Condensed Matter in 2006. In 2013, he moved to the Department of Physics, King's College London, taking a Chair in Computational Materials Science. He is now Emeritus Professor at King's College London and Senior Research Investigator in the Department of Materials at Imperial College London. He has held Visiting Professor positions at Instituto Balseiro San Carlos de Bariloche, Fraunhofer IWM Freiburg and Imperial College London.

Contents

Preface   vii

About the Author   xi

1. The Lemon Lamp   1

2. Simple Electrochemical Cells   5
    2.1   The Simplest Cell   5
    2.2   The Next Simplest Cell   7

3. Inner Potential, Work Function and Contact Potential   11
    3.1   Density Functional Theory   12
    3.2   Inner Potential Difference   15
    3.3   Contact Potential   18
    3.4   Work Function   20
    3.5   Contact Potential Again   23

4. Electrochemical Potential; Real Potential   27
    4.1   Electrochemical Potential   27
    4.2   Real Potential   29

5. Outer Electric Potential and Dipole Potential   31

6. The Bockris Point   39

7. Electron Work Function of an Electrolyte   43

8. Work Function and Real Potential of a Species in an Electrolyte   **45**

8.1 Electrochemical Potential of an Ion 45
8.2 Ion Work Function of an Electrolyte 48

9. Equilibrium and Reversible Work of the Electrochemical Cell   **53**

9.1 Standard Reaction Free Enthalpy 53
9.2 Inner Potential Difference across a Metal–Solution Interface 56
9.3 Open-Circuit Voltage — Electromotive Force 57
9.4 Single Electrode Potential 61
9.5 Electrochemical or Electromotive Force Series 63
9.6 Nernst Equation 65

10. Relative and Absolute Ion Work Function and Real Potential   **67**

10.1 Measurement of the Proton Work Function and Real Potential in Water 67
10.2 Absolute Electrode Potential 71

11. Electrode Capacitance and Electrocapillarity   **77**

11.1 Thermodynamics of the Interphase 77
   11.1.1 The case of a single phase 83
   11.1.2 The case of two phases 86
11.2 The Electrocapillary and Lippmann Equations . . . 91
   11.2.1 Derivation number one 93
   11.2.2 Derivation number two 96
11.3 Measurement 98

12. Atomistic Models of the Interphase and Interphase Capacitance   **101**

12.1 Gouy–Chapman Theory 103

13. Kinetics   **107**

13.1 Corrosion 107
13.2 First-Order Rate Equation 109
13.3 Rate Constant According to Bockris and Reddy . . 111
13.4 Velocity at the Saddle Point 115

13.5   Equilibrium Constant in Terms of Partition
Function 116

13.6   Rate Coefficient in Terms of Partition Function . . . 121

13.7   Formulation from First Principles 123

14. Single Electrode in Equilibrium and at an Overpotential   **133**

14.1   Detailed Balancing 133

14.2   Electronation 135

14.3   De-electronation 136

14.4   Further Interpretations of the Symmetry Factor:
Overpotential 138

14.5   Equilibrium Exchange Current Density 143

14.6   Interpretation of the Equilibrium Potential 144

14.7   Nernst Equation from the Point of View of
Kinetics 146

14.8   Measurement of the Exchange Current Density . . . 147

14.9   Interpretation of the Potential of Zero Charge   . . . 150

14.10 Final Remarks on the Baseness and Nobility
of Metals 152

15. Overpotential and the Butler–Volmer Equation   **155**

15.1   The Butler–Volmer Equation 155

15.2   The Limit of Small Overpotential   158

15.3   The Limit of Large Overpotential   159

16. The Evans Diagram and the Corrosion Potential   **163**

16.1   Corrosion 163

16.2   Redox Equilibrium 164

16.3   Hydrogen Evolution Reaction 168

16.4   Corrosion Potential   171

16.5   Evans Diagram 174

17. Surface Film and Pourbaix Diagram   **179**

17.1   Passivation 179

17.2   Pourbaix Diagram 181

     17.2.1   Separation into fields in the Pourbaix
diagram   181

17.3   Pourbaix Diagram for Iron 183

18. The Evils of Chloride   **187**

Appendix A Outline of the Thermodynamics of Metals and Solutions   189

A.1 Closed Systems 189
    A.1.1 State functions 189
    A.1.2 Conditions for equilibrium in closed
        systems 192
A.2 Open Systems, Chemical Potential 194
    A.2.1 Conditions for equilibrium in
        open systems 200
    A.2.2 Chemical potential in a closed system 201
A.3 Activity 203
    A.3.1 Ideal mixture 206
    A.3.2 Non-ideal mixture 208
    A.3.3 Unimolal standard state 211
    A.3.4 Chemical potential and activity of ionic
        species 214
    A.3.5 Electrolyte chemical potential
        and activity 217
    A.3.6 Measurement of mean ion activity
        coefficient 219

Appendix B List of Symbols   221

Appendix C Some Worked Problems   229

Appendix D Further Reading   261

Index   265

Chapter 1

The Lemon Lamp

We shall begin with the lemon lamp. I cut open a lemon and drive a zinc nail and a copper nail into the fruit without allowing them to touch. If I then use crocodile clips and wire to connect the two heads through a light bulb, the bulb will illuminate (see Figure 1.1). I suggest you try this at home; you could use a galvanised nail and a brass screw.

This is an electrochemical cell, as illustrated in Figure 1.2. Zinc and copper metal "electrodes" are immersed in an acid solution, that is, an aqueous solution containing hydrogen ions (protons). A chemical reaction occurs spontaneously on the left-hand side electrode:

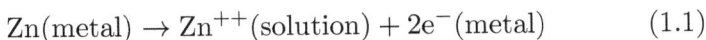

$$Zn(\text{metal}) \rightarrow Zn^{++}(\text{solution}) + 2e^{-}(\text{metal}) \qquad (1.1)$$

In this reaction, a Zn atom at the surface of the electrode detaches itself into the solution, *but it leaves behind two electrons*. In this way, it becomes a Zn^{++} "cation" free to wander in the liquid. Of course, this is *corrosion* of the zinc. The two electrons serve to reduce the charge carried by the electrode by $2e$, where, by convention, e is the fundamental constant — the charge on the proton, 1.602×10^{-19} coulomb. Because of this spontaneous reaction, this electrode is called the *electron sink*. Faraday called it the "anode". In the terminology of Bockris and Reddy (see Further Reading, Appendix D), the zinc suffers "de-electronation", also called "oxidation". It is called that because its *oxidation state* — the number of plus signs in the superscript — is increased (from zero to two).

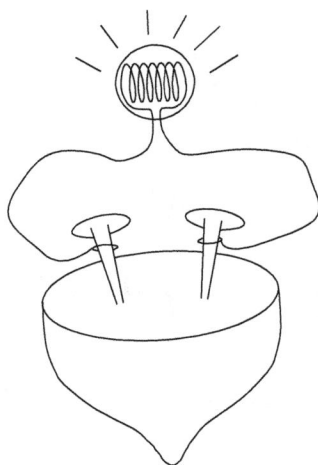

Fig. 1.1.   The lemon lamp.

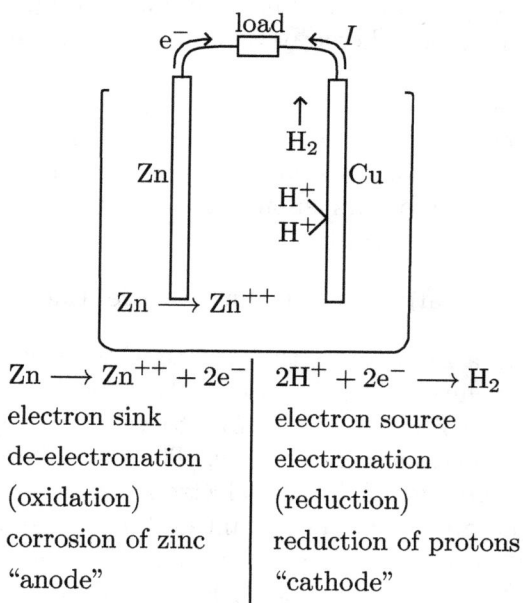

$Zn \longrightarrow Zn^{++} + 2e^-$	$2H^+ + 2e^- \longrightarrow H_2$
electron sink	electron source
de-electronation	electronation
(oxidation)	(reduction)
corrosion of zinc	reduction of protons
"anode"	"cathode"

Fig. 1.2.   Electronation and de-electronation.

Source: Adapted with permission from Bockris and Reddy (see Further Reading, Appendix D).

A second chemical reaction occurs spontaneously on the right-hand side electrode:

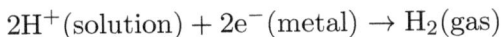

$$2H^+(\text{solution}) + 2e^-(\text{metal}) \rightarrow H_2(\text{gas})$$

Two hydrogen ions (protons) in the acid solution swipe two electrons from the electron gas in the copper and combine to form hydrogen gas which bubbles off out of the liquid and into the air above. In this way, this electrode, called the "cathode" by Faraday, is an *electron source* and the reaction is "electronation" or "reduction" since the oxidation state of the H^+ ions is reduced from plus one to zero.[1]

Because the two electrodes are connected by a wire, the electrons which are piling up on the zinc will repel each other because of the Coulomb force and will flow down the wire to the copper where they can serve the electronation reaction. In the process, they can light up the bulb. Note that even if the wires are at open circuit so that electrons cannot flow, each reaction will proceed to some limited extent until the charge accumulating on the metals prevents any further reaction. A general conclusion is that if a metal is immersed in water containing some ions, the metal may become slightly charged.

---

[1]Note that you can easily tell an oxidation from a reduction reaction because in oxidation, the electrons appear on the right of the arrow, while in reduction, they appear on the left.

Chapter 2

Simple Electrochemical Cells

2.1   The Simplest Cell

The simplest electrochemical cell is illustrated in Figure 2.1. Two electrodes made of metal M_1 and metal M_2 are immersed in a solution (or "electrolyte"), S. They are connected so as to close the circuit through a battery or voltmeter. In the lemon lamp, I glossed over the question of what metal the wires are made of. This is important all the same since the junction between dissimilar metals sets up an electric potential difference (see Section 3.4) and this will contribute to the voltage at the voltmeter. In this simplest cell, we take it that the wire is made of the same metal as one of the electrodes, say, M_1. Because the wires on either side of the voltmeter (or battery) may be at different electric potentials, we indicate this in the cartoon by including a contact between wire of metal M_1' and electrode of metal M_2.

This cell can be illustrated more symbolically as in Figure 2.2. This places the elements of the cell in line from left to right.

It is conventional to denote such a cell in a formulaic way as follows, sometimes called the "cell diagram":

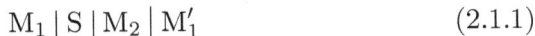

$$M_1 \,|\, S \,|\, M_2 \,|\, M_1' \qquad\qquad (2.1.1)$$

The vertical bars indicate interfaces between the phases in contact. We assume that the electric potential is uniform within each phase, but there may be an electric potential difference across each interface. The convention is to put the anode to the left where the reaction is de-electronation and the cathode on the right where the reaction is electronation (reduction).

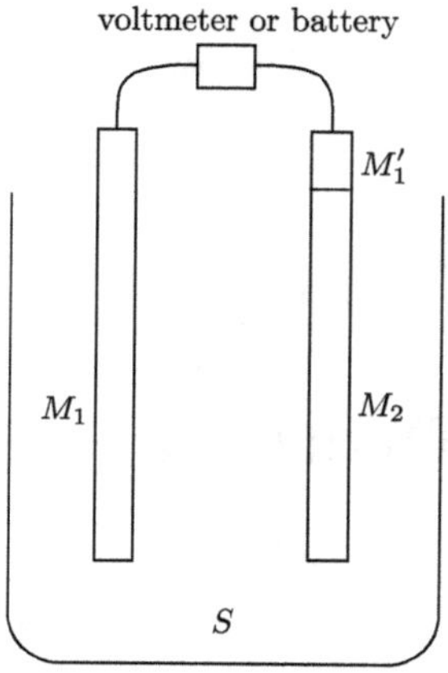

voltmeter or battery

Fig. 2.1.   The simplest electrochemical cell.

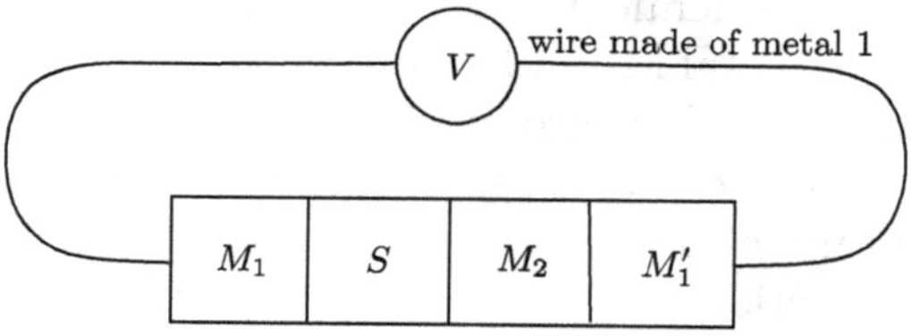

Fig. 2.2.   Symbolic cell illustration.
Source: Adapted with permission from Bockris and Reddy (see Further Reading, Appendix D).

We anticipate by our experience with the lemon lamp that an electric potential difference (PD) will be set up between the terminals of the voltmeter. Since we assume that the electric potential is uniform within each phase, there may then be an electric potential drop across each interface and the sum of these will make up the total PD across the voltmeter. This effect can be summarised in a *third* cartoon representation of the cell, namely the ladder diagram of Figure 2.3.

We concentrate for the present on the case where the circuit is closed through a high impedance voltmeter as we are interested in the electric potential difference that is set up by creating the sequence of interfaces. But keep in the back of your mind the thought that a battery in place of the voltmeter can be used to modify the PD. For example, if the zinc electrode is made more positive, it will drive the

electric potential

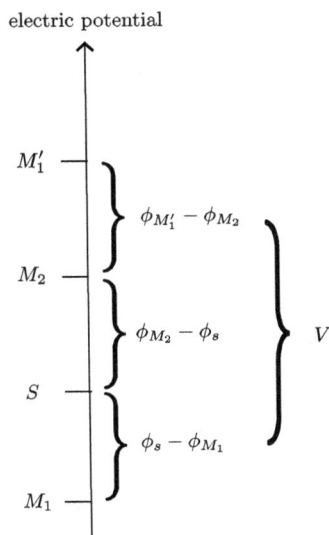

Fig. 2.3. Electric potential "ladder" diagram showing the potential differences, PD, across each of the interfaces represented by a bar, "|" in Equation (2.1.1). Electric potentials, ϕ, are labelled with subscripts to indicate the phase, or substance, in each of the boxes in Figure 2.2.

Source: Adapted with permission from Bockris and Reddy (see Further Reading, Appendix D).

reaction (1.1) to the right and increase the tendency of the zinc to corrode. Hence the metallurgical dictum: "anodes are positive and they corrode."

It is very important to appreciate what can and what cannot be measured. Clearly, we can measure the total PD across the voltmeter, but thus far, we have not established that we can measure any one of the electric potential drops across an interface. Generally, we cannot because any such measurement necessarily introduces new electrodes and probes and new PDs that conspire to add to the one we're trying to measure. As an exact statement, we may assert that *it is not possible to measure a single electrode potential*.

2.2 The Next Simplest Cell

Immediately, I have a counter response to that assertion. Figure 2.4 shows the *next* simplest electrochemical cell.

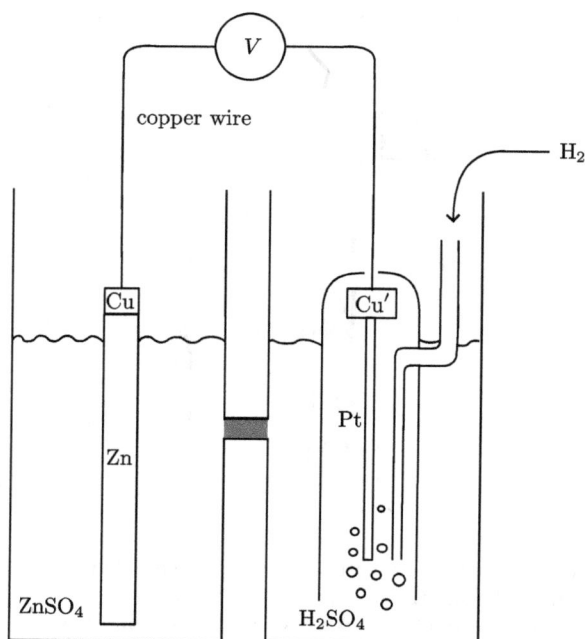

Fig. 2.4. A zinc/zinc sulphate half cell connected to a hydrogen electrode. The boxes labelled Cu and Cu′ are to remind us of the contact between the copper leads and the electrodes. The half cell on the right is made from a wire of "platinised" platinum immersed in sulphuric acid around which hydrogen gas is bubbled. If this is a *standard* hydrogen electrode (SHE), then the concentration of the acid is such that the "activity" of hydrogen ions (protons) is one and the pressure of the hydrogen gas is one bar (see Appendix A). The grey shaded link between the two half cells represents a "salt bridge", allowing current to flow between the two electrolytes while not allowing them to mix. This arrangement, unlike in the simplest cell, permits us to have the two electrodes immersed in different solutions.

In this figure, the single electrodes are separated by a narrow tube stuffed with some salt that allows ions to flow and complete the circuit but enables each electrode to be immersed in a different electrolyte. On the left is a Zn electrode immersed in a solution of zinc sulphate and so there can be an equilibrium of the reaction

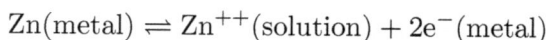

$$Zn(metal) \rightleftharpoons Zn^{++}(solution) + 2e^-(metal)$$

The other cell is the famous "standard hydrogen electrode" (SHE) which consists of a platinum strip surrounded by hydrogen gas at

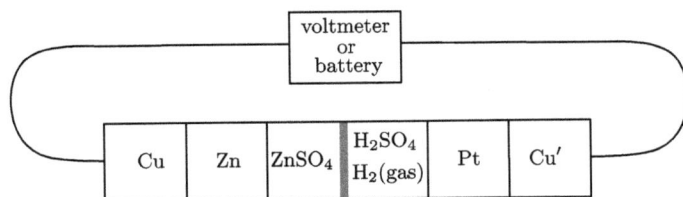

Fig. 2.5. In-line rendering of the cell of Figure 2.4.

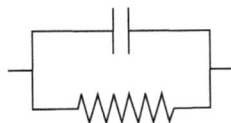

Fig. 2.6. A capacitor and resistor in parallel — model for the electrode/electrolyte interface.

exactly one bar pressure immersed in a solution of sulphuric acid at such a strength that the hydrogen ion activity (see Appendix A) is exactly one. The zinc and platinum electrodes are connected through copper wires to a voltmeter or battery. Figure 2.5 shows the cell in the in-line cartoon form.

This set up does permit the measurement of a single electrode potential because the SHE is a so-called "non-polarisable", *reversible* single electrode. This means that its single electrode potential is both known and unchanging as a result of it being connected to a second single electrode to form an electrochemical cell. A single electrode can be modelled in a first approximation as a capacitor and resistor connected in parallel, see Figure 2.6. If the resistance is zero, then the electrode is said to be "non-polarisable" so that if we try to change the electric potential on the electrode, current will leak out into the electrolyte and we find that the electric potential cannot be altered. Conversely, if the resistance is infinite, then the interface acts as a capacitor and is "ideally polarisable" (see Chapters 12 and 15), and changes in electric potential are accommodated by charge accumulation across the interface. Therefore, since all the electric potential drops are known except one, that at the Zn|S interface, then by varying the concentration of zinc sulphate and observing the voltmeter, the single electrode potential may be inferred. This is only possible if the Zn|S interface is ideally polarisable.

Chapter 3

Inner Potential, Work Function and Contact Potential

The surface of a metal has features absent in the bulk; in particular, since the electrons spill out into the vacuum, there will exist an electric dipole pointing normal to the surface from the vacuum into the metal. Similarly, for pure water, it has been established experimentally that water molecules at the surface are oriented with their oxygen atoms pointing into the vacuum; hence as for a metal surface, there is an electric dipole that points from the vacuum towards the water; the dipole potential at the surface of water is estimated from experiment to be 20 mV (see Section 10.1 and Problem 10.1). If a metal is immersed in an electrolyte, it may become charged. The liquid close to the metal surface is bound to have a molecular structure that is different from the bulk. Not only are the water molecules differently oriented near the surface anyway, but the additional charge will further modify the structure into what is called, loosely, the *double layer*. Pictures of the double layer are associated with the authors Helmholtz, Gouy and Chapman, and Stern (see Chapters 11 and 12). A common feature of the surfaces of both the metal and the electrolyte is the appearance of surface *electric dipoles*. After the interface between a metal and an electrolyte is established, these dipoles will persist and the narrow region spanning the interface on either side is often referred to in the literature as the "interphase". The purpose of the following few sections is to dissect the interphases as indicated by a vertical bar in (2.1.1) in a semi-quantitative manner.

11

Now there needs to be some remark about the word "potential" which has two meanings in all the books and literature. I have carefully used "electric potential" where convenient to describe just that: a quantity calibrated in volts or joules per coulomb in SI units. Authors persistently use "potential" when they mean potential energy whose SI units are joules. It is not really possible to write either "electric potential" or "potential energy" all the time without becoming prolix. And indeed quantities called chemical potential or inner potential are so common that we must use them — although the former is in joules and latter in volts. The most egregious villains are in density functional theory: very confusing is Lang and Kohn (see Further Reading in Appendix D) on the work function, whose ϕ is called a potential and has the usual symbol of electric potential but actually is a potential energy. At best I can warn you to be careful and hope that ambiguity doesn't creep into this book. Where possible, we will use lower case Greek ϕ, ψ and χ for electric potentials and U and V for potential energy. But we will use V for voltage. For example, if the electric potential somewhere is ϕ, then the potential energy of a charged object at the same place is $q\phi$, where q is the charge of the object. If the object is an electron, then the potential energy is $-e\phi$ (measured in electron volts, eV). Finally, recall that *all* instances of either electric potential or potential energy *must* be referenced to some arbitrary and agreed zero (see Section 10.2): it is always safest to mention potential differences rather than absolute potentials. For example, in electrostatics, we often take the electric potential as relative to a point at infinity out of range of all charges. In solid state physics, if we treat only a bulk solid with no surfaces but periodic boundary conditions, then we choose a potential energy zero with reference to some mean energy within the solid (for example, the average Hartree plus external potential energies).

3.1 Density Functional Theory

This is the only section that contains any quantum mechanics, so if you are unfamiliar, but maybe have some understanding of a valence band in a metal or semiconductor, then look at Figure 3.1 and Equation (3.1.1) and move to Section 3.2.

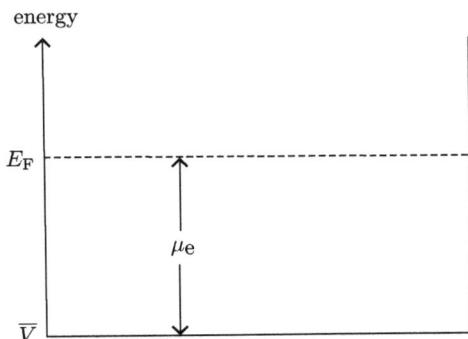

Fig. 3.1. Energy levels in a metal.

Figure 3.1 shows a sketch of the energy levels in a metal. The lower horizontal line indicates the average potential energy, $\overline{V} = -e\bar{\phi}$, of all the electrons in the metal. Because electrons are fermions, they occupy successively higher energy levels such that the energy of the highest occupied state is the Fermi energy, E_{F}. The bulk of the metal, far from its surfaces, must have its electronic structure exactly as described using periodic boundary conditions, and as long as the metal is not charged, we may describe the electronic levels using bulk density functional theory (DFT). A postulate of the DFT is that the total energy is formed of three terms, each a functional only of the ground state electron density, $n(\mathbf{r})$ (number of electrons per unit volume):

$$E_{\mathrm{tot}}[n] = T[n] + V_{\mathrm{en}}[n] + V_{\mathrm{ee}}[n]$$

which are the expectation values of the the kinetic energy, electron–nucleus and electron–electron operators. The number of electrons is

$$N = \int n(\mathbf{r}) \mathrm{d}\,\mathbf{r}$$

Often we pull out the universal functional,

$$F[n] = \left\langle \Psi \left| \hat{T} + \hat{V}_{\mathrm{ee}} \right| \Psi \right\rangle$$

(Ψ is the ground state many-electron wavefunction) and write,

$$E_{\mathrm{tot}} = F + V_{\mathrm{en}} = F + \int n(\mathbf{r}) v_{\mathrm{ext}} \mathrm{d}\,\mathbf{r}$$

where v_{ext} is the external potential (i.e., that due to the nuclei). Note that v_{ext} is a potential energy, not an electric potential. In addition, we pull out from F the classical Coulomb self-energy of the inhomogeneous electron gas, namely the Hartree energy,

$$F = E_{\mathrm{H}} + G$$

which defines the universal functional, G, which contains all the contributions to the total energy from exchange and correlation, and the interacting part of the kinetic energy. The second Hohenberg–Kohn principle states that the ground state density is that which minimises the total energy: this requires a minimisation constrained to conserve the number, N, of electrons,

$$\frac{\delta}{\delta n}\left\{E_{\mathrm{tot}}[n] - \mu\left(\int n(\mathbf{r})\mathrm{d}\mathbf{r} - N\right)\right\} = 0$$

in which μ is a Lagrange multiplier. This leads to the Euler–Lagrange equation,

$$\frac{\delta E_{\mathrm{tot}}}{\delta n} = v_{\mathrm{ext}} + V_{\mathrm{H}} + \frac{\delta G}{\delta n}$$
$$= \mu$$

where, the $\delta/\delta n$ being a functional derivative with respect to the function $n(\mathbf{r})$ (see Problem 3.2),

$$V_{\mathrm{H}} = \frac{\delta E_{\mathrm{H}}}{\delta n}$$

is the "Hartree potential" (again, actually a potential energy) and

$$\mu = v_{\mathrm{ext}} + V_{\mathrm{H}} + \frac{\delta G}{\delta n}$$
$$= \left(\frac{\partial E_{\mathrm{tot}}}{\partial N}\right)_{v_{\mathrm{ext}}} \tag{3.1.1}$$

is the zero temperature chemical potential, or variation of the total energy with respect to an infinitesimal change in electron number at fixed external potential (see Problem 3.3).

In DFT, the zero of energy is the energy of all the electrons and nuclei infinitely separated (the nuclei are not disassembled) so E_{tot} and μ are both referred to the DFT energy zero. On the other hand, if we make the choice that the zero of energy is the average potential energy of the electrons in the metal, then we have,

$$\overline{V} = \overline{v_{\text{ext}}} + \overline{V_{\text{H}}} = 0$$

and we may define

$$\mu_{\text{e}} = \mu - \overline{V} = \frac{\overline{\delta G}}{\delta n} \qquad (3.1.2)$$

as the "chemical potential of the electron in the metal". The overbar denotes a spatial average, say, over a unit cell of the metal crystal. We shall adhere to the convention that we call μ_{e} the "chemical potential" and that it is always measured from the "mean inner potential energy", $\overline{V}$.

3.2 Inner Potential Difference

Consider two metals A and B (Figure 3.2). The Fermi energies of electrons in metals A and B will not be the same; therefore, if I put specimens of the two metals into electrical contact, as, for example, at the $M_2|M_1'$ interface in (2.1.1) (or if I connect them with a wire), then if, say, $\mu_A > \mu_B$, relative to the DFT zero of energy, electrons will tend to flow from metal A to metal B. However in doing so, each metal will become charged, resulting in a shift in the potential energy, $\overline{V}$, of the electrons, and this will introduce an electric potential difference, $\Delta\phi = \phi_B - \phi_A$, between interior points in the two metals. As the electric potential in metal A becomes increasingly positive and in

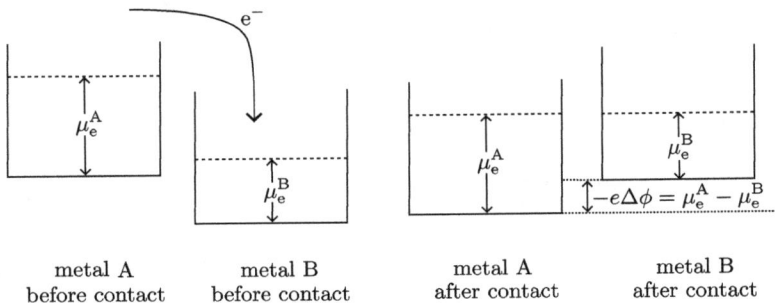

Fig. 3.2.   Band structures of two metals before and after electrical contact.

metal B negative, transfer will then be resisted until the Fermi levels align by virtue of the shift in the potential energies in each metal. This is illustrated in Figure 3.2. It is clear by construction that energy levels are shifted so that $-e\Delta\phi = \mu_e^A - \mu_e^B$. We therefore have, in equilibrium[1]

$$\mu_e^A - e\phi_A = \mu_e^B - e\phi_B \tag{3.2.1}$$

where $-e$ is the charge on the electron. The electric potential across the metal junction is sketched in Figure 3.3. Because a metal is a conductor, all the excess charge resides at the interface.

The charge averaged over the area of the interface forms a one-dimensional distribution of charge density, $\delta\rho(x)$, if x is the Cartesian coordinate normal to the interface. This generates a dipole moment per unit area, $\mathbf{p}$. If we treat the double layer as a capacitor, as illustrated in Figure 3.3, then the electric field is zero to either side and is constant and equal to σ/ϵ_0 in the gap of width d, where σ is the charge per unit area. Therefore, the electric potential drop across the interface is

$$\phi_B - \phi_A = -\frac{1}{\epsilon_0}\sigma d = \frac{1}{\epsilon_0}p \tag{3.2.2}$$

in SI units. (Note that p is negative since $\mathbf{p}$ points in the negative x-direction.) This is the "inner potential difference" or "Galvani

---

[1]Suppose N electrons are transferred from metal A to metal B. If we refer the total energy to the average potential energy, $\overline{V}$, in each metal and hence define $\overline{E}(N) = E_{\text{tot}} - \overline{V}$ as the total energy once N electrons have been added to the metal, then the total energy of the assembly is $\overline{E}_A(-N) + \overline{E}_B(N)$ plus the energy stored in the capacitor that is created by the charge dipole layer at the interface (Figure 3.3). Then if C is the capacitance,

$$\overline{E}_{A+B}(N) = \overline{E}_A(-N) + \overline{E}_B(N) + \frac{1}{2}\frac{e^2N^2}{C}$$

$$= \overline{E}_A(0) - N\frac{\partial\overline{E}_A}{\partial N} + \overline{E}_B(0) + N\frac{\partial\overline{E}_B}{\partial N} + \frac{1}{2}\frac{e^2N^2}{C} \quad \text{(to first order)}$$

$$= \overline{E}_A(0) - N\mu_e^A + \overline{E}_B(0) + N\mu_e^B + \frac{1}{2}\frac{e^2N^2}{C}$$

If this is minimised with respect to N, we obtain (3.2.1) after noting that by comparison with a flat plate capacitor, $-eN/C = \phi_B - \phi_A$.

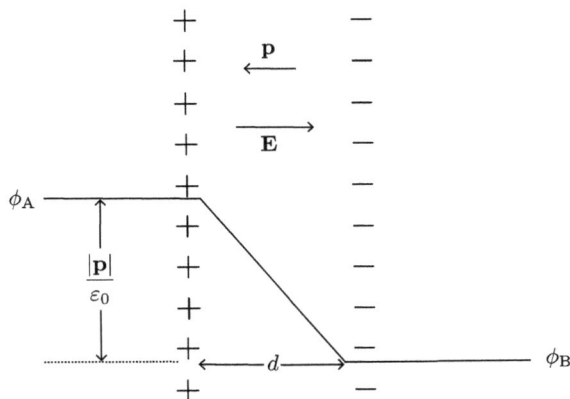

Fig. 3.3.   The electric potential across a junction between two metals. The charge on each metal is concentrated at the surface of each.

potential difference". Following conventional, if cumbersome, nota-tion, we must write this as[2]

$$\phi_B - \phi_A = {}^B\Delta^A\phi \qquad (3.2.3)$$

The result (3.2.2) is much more general that it seems. As long as the one-dimensional charge density, $\delta\rho(x)$, is zero on both sides far from the interface and that it integrates to zero between these limits so that there is a dipole but the interface is not charged, then the work done in carrying an electron from left to right across the dipole layer is

$$W_{\text{dipole}} = -e\left(\phi_B - \phi_A\right) = -\frac{1}{\epsilon_0}ep \qquad (3.2.4)$$

Briefly, the reasoning is this. The perturbation in charge density, $\delta\rho$, that creates the dipole layer is zero far to the left and far to the right of the interface and so is zero for all $x < a$ and $x > b$, where $x = a$ and $x = b$ are points in the interior of metals A and B, respectively. The difference in potential, $\phi_B - \phi_A$, is the work done in taking a

---

[2]Not only is it ugly, it is awkward and confusing, but it can be useful to shorten formulas when we add lots of PDs together. I am following Bockris and Reddy (see Further Reading in Appendix D) and West (see Further Reading in Appendix D) in putting the superscripts in the same order on both sides of (3.2.3). Other authors place them the other way around. This notation is probably best avoided.

unit test charge from $x = a$ to $x = b$ and this is minus the integral of the electric field, $E(x)$, between $x = a$ and $x = b$:

$$\phi(b) - \phi(a) = -\int_a^b E\,\mathrm{d}x = -\int_a^b E\left(\frac{\mathrm{d}}{\mathrm{d}x}x\right)\mathrm{d}x$$

$$= \int_a^b x\,\frac{\mathrm{d}E}{\mathrm{d}x}\,\mathrm{d}x + [Ex]_a^b$$

$$= \frac{1}{\epsilon_0}\int_a^b x\,\delta\rho(x)\mathrm{d}x$$

$$= \frac{1}{\epsilon_0}p$$

We inserted a "one" in the first line and integrated by parts with the boundary term vanishing since there is no electric field in the interior of the metal. The third line follows from Poisson's equation:

$$\frac{\mathrm{d}E}{\mathrm{d}x} = -\frac{\mathrm{d}^2\phi}{\mathrm{d}x^2} = \frac{1}{\epsilon_0}\delta\rho(x)$$

and the integrand in the third line is by definition the dipole moment per unit volume of the charge distribution.

It must be recalled that $\phi_B - \phi_A$ *cannot be measured.* It may have been sensible to call this the "contact potential", and indeed Bockris and Reddy do so, but this phrase is reserved for a subtly different electric potential difference between two metals in contact which *is* measurable and is the subject of the following section.

3.3   Contact Potential

There is no doubt that the electric potential at a point "just outside" the surface of a metal, say 10^{-5} cm distant, is not the same as the electric potential at infinity (out of reach of all charges and fields). The reason for this is that as electrons spill out of the surface of the metal into the vacuum, they create a charge dipole layer exactly as envisaged in Figure 3.3 if metal B is replaced by vacuum. The details of the rearrangement of charge, both at the atomic level and as averaged over thousands of surface atoms, depend on the indices of the crystal face and at the macroscopic level on the state of the surface — its roughness, contamination by adsorbates, oxidation products and so on. So each actual surface of a metal specimen will have a different distribution of polarisation (bound) charge which will give rise to stray electric fields outside the metal. These particularly arise at the corners of a single crystal or as "patch fields" which are external electric field lines connecting the surfaces of different grains in a polycrystalline sample. The existence of this electric potential is proved beyond doubt by experiments first made by Sir William Thomson, Lord Kelvin. The simplest realisation is to place the faces of two metals, or different faces of two single crystals of the same metal, into the form of a capacitor so that the two plane surfaces are parallel and separated by a distance d. If there exists an electric potential difference $\Delta\psi$ between points close to each surface, then there must be an electric field, E, and surface charge densities $\pm\sigma$ such that

$$\Delta\psi = Ed = \frac{\sigma d}{\epsilon_0}; \qquad C = \frac{A\epsilon_0}{d} \qquad (3.3.1)$$

by analogy with a parallel plate capacitor of area A and capacitance C. If the surfaces are moved further apart or closer together such that d changes but $\Delta\psi$ is to remain the same, then the surface charge density must change and this will lead to a measurable current if the two metals are connected through an ammeter. This is indeed observed, and if in addition to the ammeter a battery and potentiostat are introduced into the circuit, then a bias potential may be imposed. When this bias potential is exactly equal and opposite to $\Delta\psi$, then no current will flow when d is altered and this measured bias potential provides a *measurement* of $\Delta\psi$. This is the Kelvin probe (see Problem 3.5). The bias potential that exactly opposes the contact potential is called the *compensation potential*. For two metals, A and B, the measured difference in outer electric potential, $\psi_B - \psi_A$, is called the *contact potential* difference. It is not the same as the inner potential difference (3.2.3) which is not measurable but which does contribute to the total PD measured across an electrochemical cell due to the use of different metal electrodes and electrical connectors.

3.4   Work Function

The Kelvin probe can also be used to measure the work function.
It is easy to get confused by what is meant by work function and
so here I will attempt to stay well grounded in either experiment or
the density functional theory. However, the two approaches actually
give slightly different definitions of work function. Loosely stated, the
work function of a metal is the amount of work needed to remove
an electron from the metal (as observed, for example, in thermionic
emission or the photoelectric effect). Naturally, an electron that has
the highest (kinetic) energy will be the easiest to remove, so the
electron will originally be at the Fermi surface.

(1) Kittel[3] defines the work function, W, as "...the difference in
potential energy of an electron between the vacuum level and
the Fermi level". The vacuum level, according to Kittel is "...the
energy of the electron at rest at a point sufficiently far outside
the surface so that the electrostatic image force on the electron
may be neglected — more than 100Å [10^{-6} cm] from the surface". Different crystal surfaces have different work functions. For
example, tungsten has those work functions listed in Table 3.1.
This reflects the fact stated earlier that the electric potential is
not the same just outside different faces of the same metal (see
Problem 3.5).

Table 3.1.   Some measured work functions in electron volts
of three crystallographic faces of a tungsten crystal.

Surface	Work function (eV)
(100)	4.63
(110)	5.25
(111)	4.47

(2) The definition of work function in DFT is

$$W^{\infty} = [V(\infty) + E_{\text{tot}}(N-1)] - E_{\text{tot}}(N) \tag{3.4.1}$$

---

[3]Charles Kittel, *Introduction to Solid State Physics*, 8th edn. (John Wiley,
New Jersey, 2004).

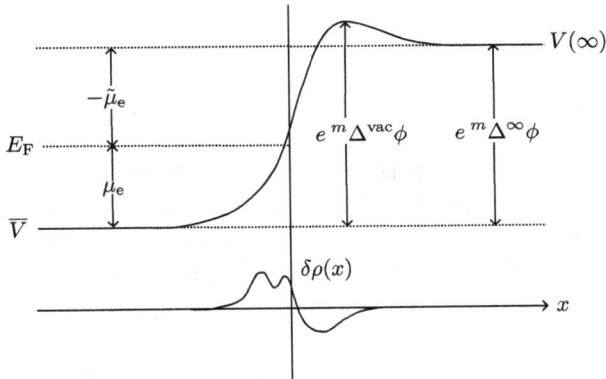

Fig. 3.4. The electric potential energy of an electron at the metal/vacuum interface. $\delta\rho(x)$ is a sketch of the possible electric charge density arising from electrons "spilling out" from the surface.

Source: Adapted with permission from N. D. Lang and W. Kohn, *Phys. Rev. B*, **1**, 4555 (1970).

In words, this is the difference in total energy between a piece of metal containing N electrons and the same piece containing $N-1$ electrons plus the potential energy of an electron at infinity. This definition can be made consistent with the first by replacing $V(\infty)$ with $V(\text{vac})$ which is the potential energy of the electron "just outside" the surface relative to the potential energy at infinity: the vacuum level.

A way to picture the description embodied in (3.4.1) is to modify Figure 3.1 to include the presence of a surface and vacuum.[4] Figure 3.4 shows the metal to the left with the quantities $\overline{V}$, E_{F} and μ_e as in Figure 3.1. To the right is the vacuum, and the potential energy of an electron is now sketched as increasing from left to right up to $V(\infty)$ as $x \to \infty$. Lang and Kohn (see Further Reading in Appendix D) show that without approximation in DFT, (3.4.1) is formally and exactly equal to

$$W^{\infty} = V(\infty) - \mu \qquad (3.4.2)$$

---

[4]By "vacuum", we really mean gas phase. In fact, in the vacuum outside a metal, there is a finite probability of finding an electron according to the Fermi–Dirac distribution. If there were not, then there would be no possibility of equilibrium between the condensed and the gas phase.

in which μ is given by (3.1.1). Lang and Kohn then add and subtract $\overline{V}$ and arrive at the formula

$$W^\infty = \left(V(\infty) - \overline{V}\right) - \left(\mu - \overline{V}\right) = e\ {}^{\mathrm{m}}\Delta^\infty\phi - \mu_{\mathrm{e}} \qquad (3.4.3\mathrm{a})$$

where ${}^{\mathrm{m}}\Delta^\infty\phi = \phi_{\mathrm{m}} - \phi_\infty$ is the difference in electric potential between the interior of the metal and charge and field-free infinity and μ_{e} is the chemical potential as defined in (3.1.2). Note that ${}^{\mathrm{m}}\Delta^\infty\phi$ is *positive* because the electric potential is more positive in the metal than outside (that's why the electrons are trapped inside). The first term in (3.4.3a) is $-e\ {}^\infty\Delta^{\mathrm{m}}\phi = e\ {}^{\mathrm{m}}\Delta^\infty\phi > 0$.

To make contact with experimental measurement and to allow for variations of work function with crystal orientation of the surface, we may replace $V(\infty)$ with $V(\mathrm{vac})$ so that

$$W = \underbrace{\left(V(\mathrm{vac}) - \overline{V}\right)}_{\text{dipole potential}} - \underbrace{\left(\mu - \overline{V}\right)}_{\mu_{\mathrm{e}}} = e\ {}^{\mathrm{m}}\Delta^{\mathrm{vac}}\phi - \mu_{\mathrm{e}} = e\chi - \mu_{\mathrm{e}} \quad (3.4.3\mathrm{b})$$

which defines the *dipole electric potential*, χ. In this way, we have replaced ${}^{\mathrm{m}}\Delta^\infty\phi$ with

$$ {}^{\mathrm{m}}\Delta^{\mathrm{vac}}\phi = \phi - \psi = \chi > 0 $$

where ψ is the electric potential at the vacuum level, known as the "outer" electric potential, or "Volta potential", and ϕ is the electric potential in the interior of the metal, known as the inner, or "Galvani" potential. Both ϕ and ψ are relative to the electric potential at infinity. The dipole (electric) potential is (3.2.2)

$$\chi = -\frac{p}{\epsilon_0} \quad (p < 0)$$

and $q\chi$ is the work done in taking a test charge, q, from the vacuum level to the interior of the metal through the dipole layer of strength p per unit area which arises from the perturbation of the charge density from its bulk distribution due to the creation of the surface (Figure 3.3). This difference in density is sketched in Figure 3.4.

Finally, note that if $\phi \equiv {}^{\mathrm{m}}\Delta^\infty\phi$ is the electric potential in the metal relative to infinity, then it is clear from Figure 3.4 that we may define a quantity by construction

$$-\tilde{\mu}_e + \mu_e = e\phi$$

from which we define the "electrochemical potential" of the electron as

$$\tilde{\mu}_e = \mu_e - e\phi \qquad (3.4.4)$$

It is then clear that since from (3.4.3a) $W^\infty = e\phi - \mu_e$, the work function at infinity is equal to minus the electrochemical potential of the electron:

$$W^\infty = -\tilde{\mu}_e$$

It is also evident that (3.2.1) states that in equilibrium the electrochemical potentials of electrons in each metal are equal, which is the usual statement of equilibrium. It is very important to note that μ, μ_e and $\tilde{\mu}_e$ are *not* different quantities. They are the same quantity but referred to different zeros of energy. This is analogous to chemical potentials in chemical thermodynamics which are only defined with respect to some standard state.

3.5 Contact Potential Again

We can now return to the experiment in which two metals are electrically connected. Figure 3.5 shows energy diagrams for metals A and B before they are connected. These are two diagrams as Figure 3.4 with energy levels, $E_F = \mu$,[5] and V for each metal referred to the DFT zero of energy. Because the Fermi energy of metal A is higher than that of metal B, we expect that when they are connected electrically by a wire, electrons will flow so as to equalise the Fermi energies. This was already demonstrated to be the equilibrium condition in Section 3.2. The condition for equilibrium is $\tilde{\mu}_e^A = \tilde{\mu}_e^B$ and this is

---

[5]This is proved by J. F. Janak, *Phys. Rev. B*, **18**, 7165 (1978); it's a sort of DFT version of Koopmans's theorem in Hartree–Fock theory.

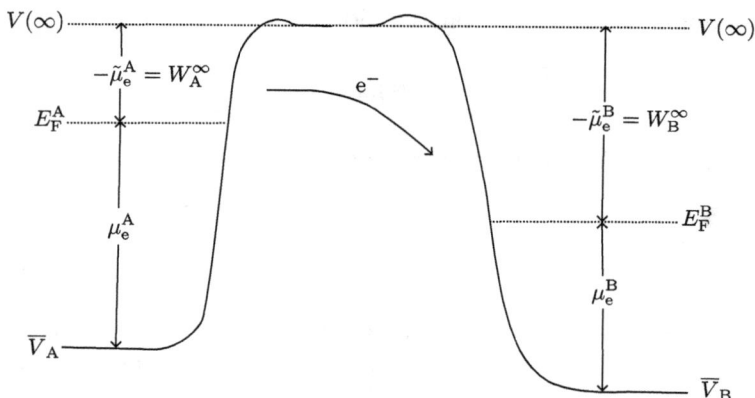

Fig. 3.5.   Electron potential energy of two metals before contact.

demonstrated in Figure 3.6, which is a cartoon of the energy levels
once the metals are electrically in contact. The solid line shows $-e$
times the electric potential in the two metals and in the vacuum
separating them. The slope in the electric potential is exactly minus
the electric field that exists in the gap between two metal surfaces in
Kelvin's experiment. Because of the transfer of electrons, both metals
are charged, and so the electric potential in each metal is uniformly
shifted. Metal A is positively charged and so its electric potential
is raised and consequently its electronic energy levels are lowered.
(It's a real nuisance that electrons are negatively charged — should
we blame Faraday? On the other hand, it does make you think —
Figures 3.5 and 3.6 would be "upside down" if we were plotting electric potential, or if you like, electrons flow uphill. The same issue
crops up in semiconductor band diagrams in textbooks.)

It is clear now that the work function is *not* minus the electrochemical potential if the metal is charged. Instead, as seen from
Figure 3.6, by construction, we have the more general formula

$$W = -e\psi - \tilde{\mu}_e \tag{3.5.1}$$

where ψ is the electric potential "just outside" the metal.

The difference between the electric potentials just outside the
two metals is $\psi_B - \psi_A$ and is called the *contact potential*; either
by construction or by subtracting (3.5.1) for two metals and noting

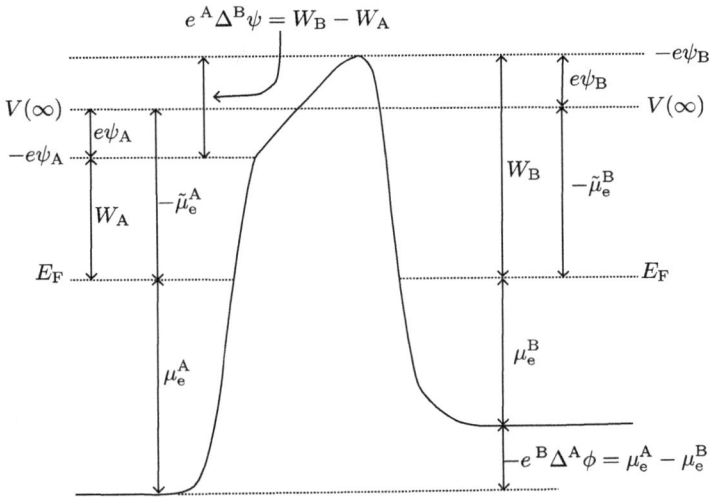

Fig. 3.6.   Electron potential energy of two metals after contact.

that their electrochemical potentials are equal when in contact, we see that

$$W_{\mathrm{B}} - W_{\mathrm{A}} = -e\left(\psi_{\mathrm{B}} - \psi_{\mathrm{A}}\right) \tag{3.5.2}$$

In words, *the difference in "outer potential" of two electrically connected metals is equal to* $-1/e$ *times the difference in work function.* We also see that the difference in inner potentials is $+1/e$ times the difference in chemical potentials, which is consistent with (3.2.1).

Chapter 4

Electrochemical Potential; Real Potential

4.1   Electrochemical Potential

The meaning of chemical potential is explained in some detail in Appendix A. At constant temperature and pressure, it is the infinitesimal change in free enthalpy (Gibbs free energy) with respect to the mole number of species i keeping the mole numbers of all other species unchanged:

$$\mu_i = \left(\frac{\partial G}{\partial n_i} \right)_{T,p,n_{j \neq i}} \tag{A2.7}$$

At constant temperature and volume, replace G with F, the (Helmholtz) free energy. Since G and F are always expressed with respect to some arbitrary zero of energy, the chemical potential is similarly undetermined to within a constant. This is usually expressed in terms of a standard chemical potential. For example, if you imagine taking a single atom or molecule from some reservoir and adding it to the body in question, then you may take the reservoir to be gas phase at standard temperature and pressure or pure condensed substance, or an atom may be taken from dilute solid solution; these choices involve different amounts of reversible work and so the chemical potentials referred to these standard states will not be the same. Just as in electrostatics, thermodynamics is about *changes* in energy; it is only necessary for consistency to keep to the same zero of energy when comparing energies. In electrolytes and solutions, the chemical potential refers to a complicated process. Adding a molecule or atom to a solution involves much complex rearrangement of its environment as the host solvent molecules and preexisting solute molecules rearrange themselves to accommodate the new species. Indeed, the work done will depend on the concentration of the species in question already present if the solute molecules are interacting. For example, we speak of the *free enthalpy of solvation* which includes the reversible work done by the solvent as its molecules create the so-called solvation shell (at constant T and p). Again, one defines a standard free enthalpy of solvation, for example, at standard temperature and pressure and at unit activity (see Section A.3) of the solute.

If the species in question is *charged*, then a polar solvent such as water will reorganise its molecular dipole moments the best to screen the electric field introduced by the charged species. Equation (A2.7) still defines the chemical potential but it is conventional to separate out from the chemical potential that part which is a result of the chemistry alone — solvation, effects of the concentration, temperature and pressure — and that part of the work done due to having to place the charge of the species i (namely $q_i = ez_i$, where z_i is called the "charge number" of the species) into the interior of a phase whose inner potential, ϕ, is different to that of the reservoir. If we agree that the species is at zero potential with respect to infinity in the reservoir, then we define the *electrochemical potential* as

$$\tilde{\mu}_i = \mu_i + Fz_i\phi \qquad (4.1.1)$$

This introduces the *Faraday constant*, F, which is the charge on a mole of protons. It is $+e$ times the Avogadro number. You may ask, how can I redefine the chemical potential (A2.7) so as to somehow exclude all effects due to the charge on the species i? After all, the structure of the solvation shell will depend on the charge. This means that the new μ_i in (4.1.1) appears to be ill-determined. The only recourse is to recognise that $\tilde{\mu}_i$ in (4.1.1) is the same quantity as defined rigorously in (A2.7) and so the new μ_i in (4.1.1) is then uniquely determined by the operation of subtracting $Fz_i\phi$ from the chemical potential. I'm sorry — this is a horrible clash of notation: I have used μ_i to mean two different things in the same section. It shan't happen again ... What I should have done is either written that (A2.7) is the chemical potential for an *uncharged* species or to have used $\tilde{\mu}_i$ in the left-hand side. To my mind, as in the case of the electron, μ and $\tilde{\mu}$ are anyway the same quantity but referred to different zeros of energy.

4.2   Real Potential

It is not possible to measure electrochemical potential because we don't really have access to a field-free infinity in the laboratory. However, we do have access to the potential just outside a surface by means of a Kelvin probe or Kenrick's experiment (see Section 10.1). Therefore, it is common to introduce an additional potential energy by replacing in Equation (4.1.1) ϕ, the inner potential, with χ, the dipole potential and to define

$$\alpha_i = \mu_i + F z_i \chi \qquad (4.2.1)$$

as the *real potential* (energy) of a species i.

You can see by comparison with (3.4.3b) and Figure 3.4 that this is the reversible work to carry a mole of species i from the vacuum level into the interior of a body. If, for some reason, we have $\psi = 0$ (say in the case of an uncharged, isotropic electrolyte, see Section 8.2), then we also have $\alpha_i = \tilde{\mu}_i$.

The point is that we now have a measurable quantity, although as I understand it only measurable after making an extra-thermodynamic assumption (see Section 10.1). From our definitions number (1) and (2) of the work functions, W^∞ and W, in Section 3.4, it should be clear that the real potential of an electron in the metal is the same as $-W$, to within a factor of the Avogadro constant.

Chapter 5

Outer Electric Potential and Dipole Potential

In electrochemistry the electrode and the solution are treated in their continuum electrostatics quite similarly. Certainly both are assumed perfect conductors in which the electric potential is uniform up to the surface or interphase and there is no electric field except at the interphase unless a current is flowing. For example, we later see that a work function can also be associated with the electrolyte, but this is a more involved quantity, firstly because the object withdrawn to the vacuum level is not usually an electron — it's an ion — and secondly because the work done depends on the state of solvation and the concentration of the ion in solution.

The inner electric potential difference (3.2.3) is called the difference in "Galvani potential", always given the symbol ϕ. Whereas under some limiting circumstances as we have seen, the Galvani potential difference can be measured (for example, for a polarisable single electrode connected to a non-polarisable reference electrode); in principle the absolute Galvani potential within the bulk of an electrode or electrolyte cannot be measured. What *can* be measured is the so-called outer electric potential, or "Volta potential" as we have seen in the Kelvin probe experiment. And so it is common to split up the Galvani potential into two terms: the Volta potential, ψ, and the remainder called the dipole potential, χ. The latter is obviously not measurable. We have

$$\phi = \psi + \chi \tag{5.1}$$

Or, at an interface, A|B (see Equation (3.2.3) and Footnote 2 of Section 3.2):

$$^{B}\Delta^{A}\phi = {}^{B}\Delta^{A}\psi + {}^{B}\Delta^{A}\chi \qquad (5.2)$$

The Volta potential difference, $\psi_B - \psi_A$, according to Bockris and Reddy is "the contribution to the potential difference across an electrified interface arising from the charges on the two phases". Imagine the process of taking a positive test charge from infinity where $\phi = 0$ and carrying it through the surface to the inside of an uncharged phase (electrode or electrolyte). We are asked to imagine first carrying the test charge to a certain point, P_B, just outside the surface where it is very close but still outside the range of the image potential. Now the image potential falls off like $1/r$, so rigorously speaking, there is no such point, but this is also implicit in Kittel's definition of work function in Section 3.4. For now, believe that this is possible and conclude that the electric potential at this point is still zero since the test charge has not encountered any electric field in its journey. *Now* suppose that the phase *is* charged. Again we imagine taking our test charge to P_B and now there will be positive or negative work done and the electric potential at P_B will not be zero. Because the point P_B is "outside the range of the image force", the work done is entirely due to moving the charge through the electric field due to the charged phase. The electric potential thus defined at point P_B is the outer electric potential, or *Volta potential*, ψ. By construction, the Volta potential of an uncharged, isotropic phase is zero (see Chapter 8).

Bockris and Reddy devised a thought experiment to clarify what is meant by the Volta potential difference across the interphase separating bulk metal (electrode) and bulk solution (electrolyte). We imagine the interphase as an assembly consisting of charged electrode, carrying a charge q_m and possessing a dipole layer at its surface where it meets the solution carrying a charge q_s. This charge is distributed near the surface of the solution since we assume that the solution is a conductor and will not support an electric field. The possibly charged layer of solution abutting the interface may be quite wide before giving way to electrolyte with bulk properties; this layer is the Gouy–Chapman or Stern double (or triple) layer (see Chapter 12), and as in the case of the metal, we expect a complex rearrangement of molecules giving rise to dipole (and presumably higher multipole) moments. The simplest assumption as in the metal is to allow a one-dimensional distribution, $\delta\rho(x)$, which integrates to zero so that the work done in taking a test charge across the double layer is equal to ϵ_0^{-1} times the dipole moment per unit area. Now Bockris and Reddy imagine separating the interphase so that all the metal is exposed to vacuum as is all the solution. They then make both halves neutral so that any dipole layer due to the charging or due to contact between the two phases vanishes. They then recharge the two phases to the original charges, q_m and q_s, and use a test charge to find the outer potentials, ψ_m and ψ_s. Then the Volta potential difference across the interphase is $^m\Delta^s\psi$.

In order to clarify the dipole potential difference across the same interphase, Bockris and Reddy devised a second thought experiment. The interphase is again dismantled and discharged, but this time the arrangement of molecules that made up the dipole, or "double", layer in the solution is reassembled on the surface of the solution. Now the separated phases exposed to vacuum are uncharged but have the dipole layers reinstated. A positive test charge is now carried from infinity to the interior of the phase where it has bulk properties, see Figure 5.1. This defines the χ-potential difference as $^m\Delta^s\chi$. In the case of the metal in Figure 5.1, it is worthwhile to compare χ_m with the work function. There are two differences. Firstly, the test charge is positive and not negative and it travels in the other direction. A subtle point we'll come back to is that the test charge is not an electron so there will be no exchange contribution to the image force. Secondly, in the second thought experiment, we bring in the test charge from infinity and not from the vacuum level. The reason for the difference in electric potential between infinity and the vacuum level of a neutral metal described in Chapter 3 is that a crystal naturally has several faces possibly with dissimilar dipole layers. Then the work required to move a test charge from infinity to the vacuum level just outside each face is affected by the electric fields produced by the changes in charge distribution at the corners of the crystal and as a result of differences in surface crystal orientation. *In all of Bockris and Reddy's thought experiments and in electrochemical textbooks, this subtle point is avoided. It is implicitly assumed that the*

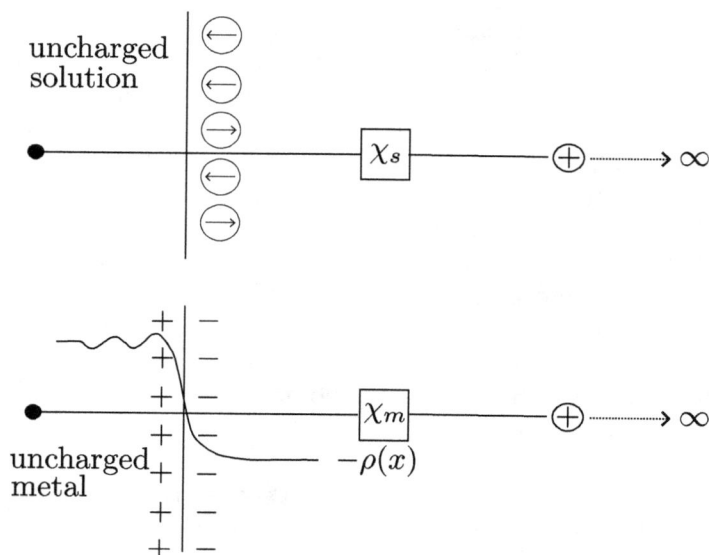

Fig. 5.1.   The work done in bringing a positive unit test charge from infinity to the interior of an *uncharged* electrolyte or electrode is the dipole potential, χ. In the case of the uncharged electrolyte, the dipole layer is represented in cartoon form as a row of positive or negative dipoles attached to the surface facing the vacuum; for the metal, the dipole layer is represented as in Figure 3.3 recognising that we expect a small "spill-out" of electrons into the vacuum as indicated in the graph of the plane-averaged electric charge density.
Source: Adapted with permission from Bockris and Reddy (see Further Reading, Appendix D).

extent of the surfaces and interfaces involved is infinite.[1] The process envisaged in Bockris and Reddy's thought experiments is illustrated in Figure 5.2. In the upper figure which illustrates the outer, Volta potential associated with the solution, ψ_s is only non-zero because the solution carries a net charge $q_s \neq 0$. If the solution were uncharged

---

[1]Imagine two thin sheets of metal joined along their flat surfaces. If these really are infinite in the y and z directions in the sheet, then to take a test charge from $-\infty$ to ∞ requires a journey through the three dipole layers and the associated work must be done; hence $\phi(-\infty) \neq \phi(\infty)$. On the other hand, if the area of metal normal to x is large but not infinite, then the test charge can be taken around the edge keeping essentially at infinity for the whole trip, in which case $\phi(-\infty) = \phi(\infty)$.

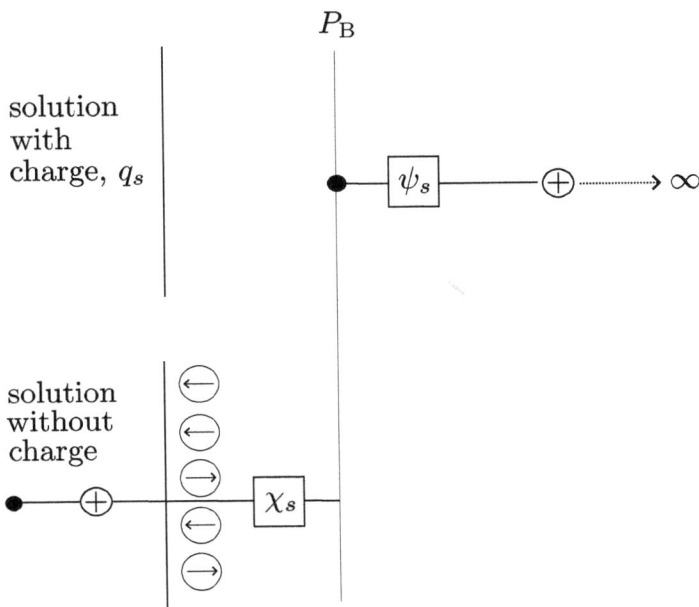

Fig. 5.2. To bring a positive unit test charge from infinity to the *Bockris point* of a *charged* electrolyte (or electrode) is its Volta potential, ψ. If the body is discharged without altering the arrangement of surface charges (the surface dipole), then the additional work needed to take the test charge from the Bockris point to deep in the interior of the phase is the dipole potential. Hence the total work done is the Galvani potential, $\phi = \psi + \chi$.
Source: Adapted with permission from Bockris and Reddy (see Further Reading in Appendix D).

then the implication is that the potential at infinity is the same as at the point P_B (which we may as well call the Bockris point); indicating that the surface is of infinite extent so there is no electric field.

Figure 5.2 illustrates the construction of the inner potential from the outer potential and the dipole potential (5.1):

$$\phi_s = \psi_s + \chi_s \tag{5.3}$$

A test charge is taken from infinity to the Bockris point; the solution is discharged while retaining its dipole double layer and the test charge is carried across the double layer into the bulk of the solution where the electric potential is ϕ_s relative to infinity. In conclusion, for any of the interfaces indicated by vertical bars in,

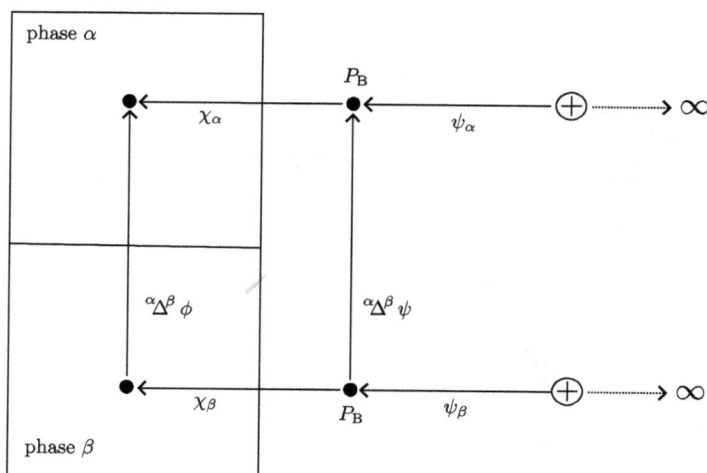

Fig. 5.3. To illustrate the inner potential difference between two phases, imagine the cycle that requires no work of taking a test charge to the Bockris point of phase α, further into the interior, across the interface into phase β and back to infinity through the surface of phase β.

Source: Adapted with permission from Fawcett (see Further Reading, Appendix D).

say, Equation (2.1.1), the electric potential difference between two contiguous phases, Figure 2.3, were it measurable, would be the difference in inner, or Galvani, potentials which is a sum of the differences of Volta and dipole potentials associated with each bulk phase.

Bockris and Reddy's thought experiments can be summarised in the diagram shown as Figure 5.3. The difference in Galvani, or inner, potential, $^{\alpha}\Delta^{\beta}\phi$, between phases α and β, be they each either electrode or solution, can only be made rigorously by reference to a surface of each phase exposed to vacuum. For each phase, a test charge is taken from infinity to the Bockris point and the work done per unit charge is the outer, or Volta, potential of each phase. If either phase is uncharged and isotropic, then this is zero for that phase. To avoid the complications of Bockris and Reddy's second thought experiment, we have to assume that all of the work against the electric field due to the charged phase has now been done so that as the test charge is now moved from P_B to the interior where the properties are uniform and bulk-like; the only work done is that against the dipole layer — this can be any charge distribution, $\delta\rho(x)$, as long as it integrates to zero between P_B and the interior point. The dipole potential is then just ϵ_0^{-1} times the dipole moment per unit area. The vertical arrows in the figure show the electric potential differences between the two Bockris points which is ${}^{\alpha}\Delta^{\beta}\psi$ and between the two interior points which is ${}^{\alpha}\Delta^{\beta}\phi$. This electric potential difference is the result of taking a test charge from the interior of the β phase to its Bockris point, on to infinity and then back to the Bockris point of the α phase and finally into the interior of that phase. This results in

$$ {}^{\alpha}\Delta^{\beta}\phi = -\chi_{\beta} - \psi_{\beta} + \psi_{\alpha} + \chi_{\alpha} $$

which confirms Equation (5.2).

Chapter 6

The Bockris Point

How far is the Bockris point from the surface of a charged metal or electrolyte? Neglecting for the moment the image potential, suppose that the phase is in the shape of a ball of radius r and carries a total charge Q. Then the outer electric potential with respect to a point at infinity at a distance s from the surface is

$$\psi_Q = \frac{1}{4\pi\epsilon_0} \frac{Q}{r+s}$$

Some authors assert that since $s \ll r$ we may write this as

$$\psi_Q = \frac{1}{4\pi\epsilon_0} \frac{Q}{r}$$

indicating that the outer potential is constant. It seems unsatisfactory to neglect s altogether since an elementary Taylor expansion results in

$$\frac{1}{r+s} = \frac{1}{r}\left(1 - \frac{s}{r}\right)$$

to first order and this is not constant but is linear in s.

The image potential seen by a test charge q at s, with respect to infinity, if the ball is conducting and grounded is[1]

$$\psi_i = -\frac{1}{4\pi\epsilon_0} \frac{1}{2} \frac{qr}{s\,(s+2r)}$$

---

[1]David J. Griffiths, *Introduction to Electrodynamics*, 4th edn. (CUP, Cambridge, 2017).

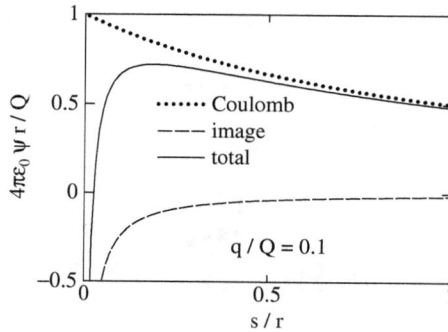

Fig. 6.1.   Electric potential relative to infinity of a point charge, q, at a distance s from the surface of a conducting ball carrying a charge, Q. The total potential is divided into Coulomb and image parts. Linear axes with $q/Q = 0.1$.

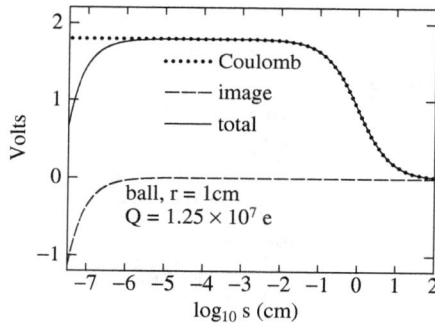

Fig. 6.2.   Electric potential relative to infinity of a point charge, q, at a distance s from the surface of a conducting ball carrying a charge, Q. The total potential is divided into Coulomb and image parts. Linear–log plot with Q/r appropriate for a ball charged to about a volt and q the elementary charge, e.

The total Volta potential, Coulomb plus image, is hence

$$\psi = \frac{1}{4\pi\epsilon_0}\left(\frac{Q}{r+s} - \frac{1}{2}\frac{qr}{s\,(s+2r)}\right)$$

$$= \frac{1}{4\pi\epsilon_0}\,\frac{Q}{r}\left(1 - \frac{s}{r} + \frac{q}{Q}\left(\frac{1}{8} - \frac{1}{4}\frac{r}{s}\right)\right) + \mathcal{O}\left(\frac{s}{r}\right)^2$$

It's not obvious that this function is constant for any $s \ll r$ and when plotted as in Figure 6.1; it clearly is not.

However, this plots the case when the test charge is as much as one-tenth of the charge on the ball. If we take a more realistic case of $q = e$, and $Q = 1.25 \times 10^7 e$ which will charge a ball of $r = 1\,\mathrm{cm}$ to about a volt, then we are rescued by making a plot against $\log s$. This is done in Figure 6.2. Now there is quite a wide plateau, and indeed these calculations are consistent with similar plots shown by Bockris and Reddy. In conclusion, the Bockris point, as asserted by Bockris, is no closer than $10^{-5}\,\mathrm{cm}$ from the surface. Infinity is seen to be at about $10\text{--}100\,\mathrm{cm}$. On the other hand, the image potential is very short ranged and as Kittel states, is negligible at about $10^{-6}\,\mathrm{cm}$ from the surface (see Section 3.4).

Note that in Figure 6.2 the test charge is taken to be a proton (or positron or antimuon). This is to acknowledge that this is a classical electrostatic calculation and the exchange contribution to the image potential is not included as it should be for an electron.

Chapter 7

Electron Work Function of an Electrolyte

In order to analyse the potential difference across an electrochemical cell, as well as ask about the work function of the metals involved, we may ask about the solutions. Sometimes the work function for an ionic species is required, and usually that is expressed in terms of the real potential, say, of species i in phase α (Section 8.2). In addition, we may ask for the work done in withdrawing an electron from an electrolyte. An elegant thought experiment by Schmickler and Santos (see Further Reading in Appendix D) illustrates this procedure and comes up with an operational definition of work function for an electrolyte. These authors imagine a metal–electrolyte interface with both phases also being exposed to vacuum. The idea is to try and transfer an electron from the solution to the metal via the vacuum. The solution contains equal numbers of moles of Fe^{++} and Fe^{+++} ions. The electron transfer is accomplished in the following five steps.

(1) Take an Fe^{++} ion out of the solution and into the vacuum just above the surface of the solution. The amount of work required is minus $\Delta_{sol}^r G(Fe^{++})$, which is the *real* free enthalpy of solvation of the Fe^{++} ion. This is a measurable real potential, and it includes the work done against the surface electric dipole belonging to the solution.

(2) Remove an electron from the Fe^{++} ion:

$$Fe^{++} \rightarrow Fe^{+++} + e^-$$

The work done is the third ionisation potential of iron, I_3.

(3) Return the Fe^{+++} ion to the solution. This gains us the real free enthalpy of solvation of the Fe^{+++} ion.

(4) Transfer the electron from the Bockris point of the electrolyte to the Bockris point of the metal. The work done is

$$-e\left(\psi_m - \psi_s\right)$$

(5) Lastly, put the electron into the metal and gain its work function, W_m.

The total work done should be zero if the electrons are in equilibrium in the assembly, so the energy budget is

$$0 = -\Delta_{sol}^r G(Fe^{++}) + I_3 + \Delta_{sol}^r G(Fe^{+++}) - e\left(\psi_m - \psi_s\right) - W_m$$

or

$$-e\left(\psi_m - \psi_s\right) = W_m - \left[\Delta_{sol}^r G(Fe^{+++}) - \Delta_{sol}^r G(Fe^{++}) + I_3\right]$$

Recall that for two metals, A and B, the difference in work functions is $-e$ times the difference in outer potentials, see Equation (3.5.2). This implies that the "work function" of the electrolyte is the sum of terms in brackets:

$$W_s = \Delta_{sol}^r G(Fe^{+++}) - \Delta_{sol}^r G(Fe^{++}) + I_3 \qquad (7.1)$$

all three quantities on the right being measurable. However, this work function depends on many things. It is the work function of the Fe^{++}/Fe^{+++} "redox couple" since we have used iron ions as the vehicle for carrying the electron away from the solution. The work function would of course be different if we used other cations (see Problem 7.1). It also depends on temperature and on the composition. It is commonplace to construct *standard* real potentials and work functions based on unit activity and standard temperature and pressure.

Chapter 8

Work Function and Real Potential of a Species in an Electrolyte

8.1   Electrochemical Potential of an Ion

The chemical potential of a pure substance as described in Equation (A2.7), is nothing other than the partial molar free enthalpy, reflecting the fact that free enthalpy is an extensive state function which is homogeneous and of first order. This allows us to derive Equation (A2.10) for the case of a substance made up of more than one component. Whereas G is only determined to within an arbitrary zero of energy, one might think that μ is independent of the choice of zero since the derivative implies a *difference* in free enthalpy which ought to cancel out the dependence on that choice. This is true for a pure substance. On the other hand, the chemical potential of a species i in a multicomponent substance as expressed in Equation (A3.1) is still undefined because we need to refer it to a "standard" chemical potential which depends only on temperature and pressure. This is because the differentiation in (A2.7) still concerns a thought experiment whereby the number of moles of component i is varied while keeping T, p and the mole numbers of all other components fixed. We are reversibly adding an infinitesimal number of moles, dn, and so the total change in free enthalpy is

$$G(n_1 \ldots, n_i + dn_i, \ldots n_N) - \big(G(n_1 \ldots n_i \ldots n_N)$$

$$+ \ \mu_i^\circ dn_i(\text{moles of } i \text{ in a reservoir})\big)$$

So the chemical potential is not independent of the zero of energy because it depends on the nature of the reservoir and the molar free enthalpy, μ_i°, of species i in the reservoir — this is the standard chemical potential. It may be the free enthalpy per mole in the pure substance solid state, or the gas phase, at standard T and p; or it may be the free enthalpy per mole in a dilute solution. In solution chemistry, we talk of Henrian and Raoultian standard states and molar and molal standard states meaning that the species is in unit concentration either in moles per litre or moles per kg of solvent. Nonetheless in the thought experiment, although we need to describe the nature of the reservoir, we do not need to consider the journey that the fraction of a mole of species takes when it leaves the reservoir and arrives in the system under study.

Everything changes when we have the complication that the species i and the system (electrode or electrolyte) may be charged. (In normal circumstances in electrochemistry, electrolytes are neutral however.) In that case, the reversible work done in carrying a charged component from infinity to the interior of a phase, as we have seen, depends on the details of the surface dipole layer and upon the Galvani potential within the bulk of the phase. It is completely unreasonable to assert that the system is in the shape of a thin specimen infinitely extended in the plane. If that object is charged, then the electric field is constant and the electric potential is linear in the distance from the sheet, so it diverges to $\phi = \pm\infty$ at infinity so there can exist no field-free infinity. The electrochemical potential is only defined for a system of finite spatial extent.

Figure 8.1 which is attributed to Roger Parsons illustrates some of the points in this chapter. The upper cartoon is to illustrate the chemical potential. We imagine the system as being bulk-like and homogeneous up to its surface. If however the phase *is* charged, then the charge will reside at the surface (since we take electrolytes to be conducting) and there will be a charged surface dipole layer. The process shown in the middle cartoon is the equivalent of the Bockris thought experiment in Figure 5.2. Now instead of a test charge, we bring in the charged species i from infinity, past the Bockris point where the Volta electric potential is ψ, through the double layer and into the interior where the Galvani potential is ϕ. This serves to *define* the electrochemical potential. The lowest cartoon is to illustrate how the electrochemical potential is divided, as in Equation (4.1.1), into

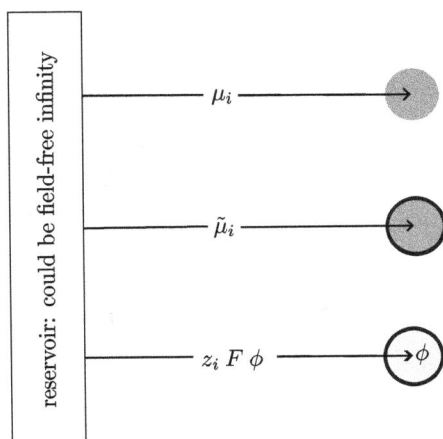

Fig. 8.1. A cartoon to illustrate the "chemical" and "electrical" components that make up the electrochemical potential, see Equation (4.1.1). In the top figure, the definition of chemical potential of an uncharged species in an uncharged system is well defined; it is the reversible work to take one atom or molecule (or alternatively an Avogadro number of such entities — a mole) from a reservoir at some temperature and pressure and place it in the bulk of the system. The reservoir may be pure solid or gas, or a dilute or concentrated solution. The centre cartoon shows the taking of a charged atom or molecule from a reservoir and carrying it into the bulk of a possibly charged system. This quantity is not necessarily even measurable, but its division into the two terms of (4.1.1) is described by imaging that we excavate out all matter from the system and leave behind only the surface charge and dipole layer. Since we assume the system is conducting, all the charge will anyway be residing at the surface. The lowest figure then illustrates the purely electric work involved in taking a charged entity from the reservoir and placing it in the centre of the vacuum surrounded by the surface charges. The electric potential within this vacuum is the inner or Galvani potential, ϕ.

Source: Adapted with permission from Fawcett (see Further Reading in Appendix D).

the electric work and the remainder. We imagine that the bulk has been excavated from the phase leaving the charged dipole layer surrounding empty space. The electric potential within the space is different from that at infinity due to the surface charge. Hence the reversible work per mole is $z_i F \phi$, where z_i is the charge number of component i and F is the Faraday constant.

Consider the case of one-to-one electrolyte, for example, NaCl dissolved in water. The electrochemical potentials of the cation and anion, respectively, are

$$\tilde{\mu}_{Na^+} = \mu_{Na} + F\phi; \quad \tilde{\mu}_{Cl^-} = \mu_{Cl} - F\phi$$

Now the Galvani potential is in principle not measurable and so the *individual electrochemical potentials of ions are not measurable*; hence they are not proper thermodynamic quantities (see Sections A.3.4–A.3.6). This is a most important observation in electrochemistry, not to be ignored. What is measurable is the electrochemical potential of NaCl in the solution, since by addition the charge term cancels and we have

$$\tilde{\mu}_{NaCl} = \tilde{\mu}_{Na^+} + \tilde{\mu}_{Cl^-} = \mu_{Na} + \mu_{Cl} = \mu_{NaCl}$$

8.2   Ion Work Function of an Electrolyte

So far we have encountered the work function of an *electron* in both a metal and an electrolyte as the reversible work to remove the electron from the interior of the phase to the vacuum level (Bockris point).

We might also ask, what is the reversible work needed to remove one mole of *ions* from an electrolyte, or solution, in which they are in a standard state of unit concentration and standard T and p, and take them to the vacuum level. For this quantity, we will use the symbol $W_i^{s,\circ}$, being the work function of component i in a solution (superscript "s") phase including a superscript, $\circ$, to make explicit that this is a standard state quantity.[1] The process under consideration is therefore,

ion in solution $\longrightarrow$ ion in gas phase at vacuum level

Once it is in gas phase, we must assume that it is accompanied by an equal number of ions of opposite charge so that the gas is neutral; in addition, we don't expect any further work to be done carrying

---

[1] My definition of ion work function is at odds with Fawcett's definition (Fawcett, see Further Reading in Appendix D) in that he replaces the vacuum level with charge-free infinity; on the other hand, my definition is consistent with that of Cheng and Sprik (see Further Reading in Appendix D).

the mole of neutral gas from the vacuum level to infinity. This means that we have a working definition of the ion work function as

$$W_i^{\mathrm{s},\circ} = \tilde{\mu}_i^{\mathrm{g},\circ} - \alpha_i^{\mathrm{s},\circ} \tag{8.2.1}$$

since the reversible work to take the ions from the interior of the phase to the vacuum level is minus the real potential (Section 4.2).

Recall that the chemical potential of a component i in a phase depends on its concentration, T and p. The superscript $\circ$ on $\alpha_i^{\mathrm{s},\circ}$ is to indicate that this is the real potential of component i when it finds itself in the solution phase at unit concentration and standard T and p.

Once the mole of ions has been deposited into its gas phase and electrically neutralised, its electrochemical and chemical potentials will be identical and we have

$$\begin{aligned} W_i^{\mathrm{s},\circ} &= \mu_i^{\mathrm{g},\circ} - \alpha_i^{\mathrm{s},\circ} && (8.2.2) \\ &= \mu_i^{\mathrm{g},\circ} - \left(\mu_i^{\mathrm{s},\circ} + z_i F \chi_{\mathrm{s}} \right) \\ &= \mu_i^{\mathrm{g},\circ} - \left(\mu_i^{\mathrm{s},\circ} + z_i F \left(\phi_{\mathrm{s}} - \psi_{\mathrm{s}} \right) \right) \\ &= \mu_i^{\mathrm{g},\circ} - \tilde{\mu}_i^{\mathrm{s},\circ} \quad \text{iff } \psi_{\mathrm{s}} = 0 && (8.2.3) \end{aligned}$$

In this way, the two definitions of Fawcett (8.2.3) and Cheng and Sprik (8.2.2) become aligned in the case that the electrolyte is uncharged and has isotropic surfaces (see Section 4.2).

For an uncharged solute species, i, and an uncharged solvent, it is customary to define the standard free enthalpy of solvation as the reversible work at constant T and p needed to take a mole of solute molecules from the gas phase at standard T and p and insert them into the solvent to achieve unit concentration at standard T and p:

$$\Delta_{\mathrm{sol}}^{\infty} G_i^{\circ} = \mu_i^{\mathrm{s},\circ} - \mu_i^{\mathrm{g},\circ}$$

I have attached a superscript ∞ to remind us that this quantity, like the work function, W^{∞}, of Equation (3.4.1) in Chapter 3, is defined in reference to infinity, not the vacuum level, since these are standard chemical potentials. Then the same formula can be applied to define a standard free enthalpy of solvation for a charged species.[2] In that case, we could write, using (4.1.1) and (8.2.3),

$$
\begin{aligned}
-\Delta_{\text{sol}}^{\infty} G_i^{\circ} &= \mu_i^{\text{g},\circ} - \mu_i^{\text{s},\circ} \\
&= \mu_i^{\text{g},\circ} - \left(\tilde{\mu}_i^{\text{s},\circ} - z_i F \phi_{\text{s}} \right) \\
&= W_i^{\text{s},\circ} + z_i F \phi_{\text{s}} \qquad \text{iff } \psi_{\text{s}} = 0 \\
&= W_i^{\text{s},\circ} + z_i F \chi_{\text{s}} \qquad \text{iff } \psi_{\text{s}} = 0 \qquad (8.2.4)
\end{aligned}
$$

Therefore, again only in the case of an uncharged and isotropic electrolyte, we have

$$
\Delta_{\text{sol}}^{\infty} G_i^{\circ} = - \left(W_i^{\text{s},\circ} + z_i F \chi_{\text{s}} \right) \qquad \text{iff } \psi_{\text{s}} = 0 \qquad (8.2.5)
$$

We have already encountered the free enthalpy of solvation in Chapter 7. There we were concerned with the *electron* work function of an electrolyte, but in order to remove an electron, we used an iron ion as the carrier. However, in that case, since the ion is charged, the correct quantity is the *real* solvation free enthalpy, which in contrast to (8.2.4) is

$$
\begin{aligned}
\Delta_{\text{sol}}^{\text{r}} G_i^{\circ} &= \alpha_i^{\text{s},\circ} - \mu_i^{\text{g},\circ} \\
&= -W_i^{\text{s},\circ} \qquad \text{iff } \psi_{\text{s}} = 0 \qquad (8.2.6)
\end{aligned}
$$

Finally, since the quantities defined in this section are all standard free enthalpies, it is unsurprising that in the case that the solution is not at unit concentration, we use (A3.1)

$$
\alpha_i^{\text{s}} = \alpha_i^{\text{s},\circ} + RT \ln a_i
$$

$$
\Delta_{\text{sol}}^{\text{r}} G_i = \Delta_{\text{sol}}^{\text{r}} G_i^{\circ} + RT \ln a_i
$$

where the activity is defined with respect to unit concentration. I don't think that people ever write $W_i^{\text{s},\circ} = W_i^{\text{s}} + RT \ln a_i$ which is probably why they don't put a superscript $\circ$ on the ion work function.

---

[2]I have difficulty with this because to me μ_i for a charged species is not physically motivated as I pointed out in comments after Equation (4.1.1). To me, $\tilde{\mu}_i$ is well defined and μ_i is just the *remainder* after the electric work has been subtracted.

It is worthwhile to bear in mind when reading the literature that in general authors will assume an uncharged, isotropic electrolyte. In that case, it is clear from Equations (8.2.2), (8.2.3) and (8.2.6) that

$$\Delta_{\text{sol}}^{\text{r}} G_i^{\circ} \equiv -W_i^{\text{s},\circ}$$

and

$$\tilde{\mu}_i^{\text{g},\circ} \equiv \alpha_i^{\text{s},\circ}$$

are *identities* and so confusion may arise since neither $W_i^{\text{s},\circ}$ nor the real potential actually needed to be invented at all! On the other hand, because they are implicitly referred to the vacuum level and not to field free infinity, these are in principle *measurable* quantities.

Chapter 9

Equilibrium and Reversible Work of the Electrochemical Cell

Possibly the most pressing questions in electrochemistry are, which metals are more likely to corrode; and how much energy can I get out of an electrochemical cell?

9.1 Standard Reaction Free Enthalpy

Suppose we are interested in a chemical reaction[1]

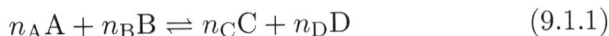

$$n_A A + n_B B \rightleftharpoons n_C C + n_D D \qquad (9.1.1)$$

in which n_A moles of a substance A and n_B moles of a substance B react to form n_C moles of a substance C and n_D moles of a substance D. On the left are *reactants*, and on the right are *products*. If this reaction occurs spontaneously, then the change in free enthalpy

---

[1]In Chapter 1, I wrote "reactants" and "products" separated by an arrow to indicate what is reacting to become what, but it is more indicative to use the double harpoons in chemistry to emphasise that usually a reaction may proceed in either direction, depending on the concentrations of the chemicals and on laboratory conditions, especially temperature and pressure. It is important to be aware that *in equilibrium* the forward and backward rates of reaction are equal (see Section 14.1), and so on a macroscopic scale, nothing appears to be happening.

is, in view of (A2.10),

$$\Delta G = (n_C \mu_C + n_D \mu_D) - (n_A \mu_A + n_B \mu_B) < 0$$

If we use (A3.1) to expand the chemical potentials, we get

$$\Delta G = (n_C \mu_C^\circ + n_C RT \ln a_C + n_D \mu_D^\circ + n_D RT \ln a_D)$$
$$- (n_A \mu_A^\circ + n_A RT \ln a_A + n_B \mu_B^\circ + n_B RT \ln a_B)$$

Gathering terms, we have

$$\Delta G = (n_C \mu_C^\circ + n_D \mu_D^\circ) - (n_A \mu_A^\circ + n_B \mu_B^\circ)$$
$$+ (n_C RT \ln a_C + n_D RT \ln a_D) - (n_A RT \ln a_A + n_B RT \ln a_B)$$
$$= \Delta G^\circ + RT \ln \frac{a_C^{n_C} a_D^{n_D}}{a_A^{n_A} a_B^{n_B}}$$
$$= \Delta G^\circ + RT \ln K$$

This serves to define for us the *standard free enthalpy change* of the reaction (9.1.1)

$$\Delta G^\circ = (n_C \mu_C^\circ + n_D \mu_D^\circ) - (n_A \mu_A^\circ + n_B \mu_B^\circ) \qquad (9.1.2)$$

and the *equilibrium constant*

$$K = \frac{a_C^{n_C} a_D^{n_D}}{a_A^{n_A} a_B^{n_B}} \qquad (9.1.3)$$

It is easy enough to generalise (9.1.2) and (9.1.3) to the case of any number of reactants and products using summation and product signs. If the reaction (9.1.1) is in equilibrium, that is, the rate of reaction to the right is equal to the rate of reaction to the left, then $\Delta G = 0$ and we have

$$K = e^{-\Delta G^\circ / RT}$$

This is called the "van 't Hoff isotherm" and given the standard free enthalpy of the reaction allows us to calculate the equilibrium activities of the component species. Observe that the equilibrium constant is not constant at all — it depends exponentially on the temperature.

I have used the generic superscript o to indicate the standard state. But once we specify a reaction I can use, for example, • if the component is in the pure substance standard state or ⊖ if the substance is in the one bar gaseous state, and I can mix these in the formula pair (9.1.2) and (9.1.3) as long as I use the same standard for μ° and the activity for a particular species.

In electrochemistry, and in corrosion, in particular, we are interested in the oxidation of metals. For a metal M, the reaction of interest is

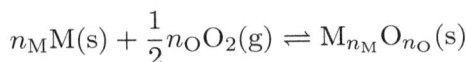

$$n_M M(s) + \frac{1}{2} n_O O_2(g) \rightleftharpoons M_{n_M} O_{n_O}(s)$$

where "s" stands for solid phase and "g" stands for gas phase. The standard free enthalpy of formation of the oxide is (9.1.2)

$$\Delta_f G^\circ = \mu^\bullet_{M_{n_M} O_{n_O}} - n_M \mu^\bullet_M - \frac{1}{2} n_O \mu^\ominus_{O_2} \qquad (9.1.4)$$

We are free to choose zeros of energy such that the free enthalpy of a pure substance in its standard state is zero. Some measured standard free enthalpies of formation (in kilojoule per mole) with respect to the pure reactants are shown in Table 9.1.

All these are negative, which reflects humankind's struggle through the copper, bronze and iron ages to win metals from their ores and the fight to prevent these returning to the oxide — that is, corroding and oxidising. The only metal found in its noble state in the earth is gold (apart from some meteoric iron) which is why it is highly prized. At the end of civilisation when Gaia returns to her natural owners, there will be no trace of the metals that were won from

Table 9.1.   Standard free enthalpies of formation of some oxides (kJ mol^{-1}).

MgO	TiO$_2$	NiO	Al$_2$O$_3$	FeO	Fe$_2$O$_3$	Fe$_3$O$_4$	ZnO	H$_2$O
−300	−90	−245	−1700	−270	−820	−1120	−350	−285

Au$_2$O$_3$	MnO	Mn$_2$O	HgO	SiO$_2$	AgO	V$_2$O$_3$	RuO$_2$
−2	−360	−460	−60	−860	−10	−65	−250

their ores during the disastrous evolutionary experiment of combining a large fore-brain with opposable thumbs (with acknowledgement to Amory Lovins).

9.2 Inner Potential Difference across a Metal–Solution Interface

In Section 3.2, we calculated the inner potential difference across a metal–metal interface: it's the difference in Fermi levels. Now we are in a position to do the same at an electrode–electrolyte interface. Consider the "half cell" in which Fe metal is immersed in a solution of Fe^{++} ions. The following chemical reaction will occur:

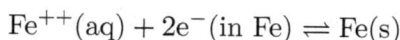

$$Fe^{++}(aq) + 2e^-(in\ Fe) \rightleftharpoons Fe(s)$$

"aq" means in aqueous solution. The free enthalpy change per mole is

$$\Delta G = \mu_{Fe} - \tilde{\mu}_{Fe^{++}} - 2\tilde{\mu}_e$$
$$= (\mu_{Fe}^{\bullet} + RT\ln a_{Fe}) - (\mu_{Fe^{++}}^{\square} + RT\ln h_{Fe^{++}} + 2F\phi_s)$$
$$- 2\left(2\mu_e^{Fe} - F\phi_m\right)$$

$h_{Fe^{++}}$ is the practical Henrian "working" activity of iron ions (see Section A.3.5, Equation (A3.26a)) and the superscript $^{\square}$ indicates the unimolal (one mole per kg solvent) standard state.[2] μ_e^{Fe} is the (standard) chemical potential of the electron in the metal, $\tilde{\mu}_e = \mu_e - F\phi_m$.[3] Having immersed the metal in the electrolyte, when the half

---

[2]If you can't be bothered with the details of how the "activity" is worked out, then at a first reading of this and what follows in Chapters 9 and 10, it is quite adequate to take the activity, h, of some species as its concentration in moles per kg solvent and its standard chemical potential, $\mu^{\square}$, as its value at $1\,mol\,kg^{-1}$.

[3]Compare this with (3.4.4). I will not distinguish in the notation between free enthalpy *per mole* and free enthalpy *per particle*. It will be obvious from the context.

cell comes to equilibrium, $\Delta G = 0$, and we have

$$2F\left(\phi_{\mathrm{m}} - \phi_{\mathrm{s}}\right) = \left(\mu_{\mathrm{Fe}^{++}}^{\square} - \mu_{\mathrm{Fe}}^{\bullet} + 2\mu_{\mathrm{e}}^{\mathrm{Fe}}\right) + RT \ln \frac{h_{\mathrm{Fe}^{++}}}{a_{\mathrm{Fe}}}$$

$$= \mu_{\mathrm{Fe}^{++}}^{\square} + 2\mu_{\mathrm{e}}^{\mathrm{Fe}} + RT \ln h_{\mathrm{Fe}^{++}} \qquad (9.2.1)$$

The second line follows because the standard free enthalpy of the pure solid iron is zero and its activity is one. Equation (9.2.1) should allow us to deduce $\Delta\phi_{\mathrm{eq}} = {}^{\mathrm{m}}\Delta^{\mathrm{s}}\phi$ (see ahead to Section 14.1) from the concentration of iron ions in the electrolyte. However, this can only be known relative to some other half cell: the absolute value of $\mu_{\mathrm{Fe}^{++}}^{\square}$ can only be measured relative to another ion as the electrochemical potential of a charged ion is not measurable (Section 8.1). Moreover, the activity of an individual ion is also not measurable; we know the concentration of ions in moles per kg of water, but we don't know the activity coefficient — we have to assume a mean ion activity coefficient (A3.19). Relative to the hydrogen ion, the standard (unimolal) chemical potential of the Fe^{++} ion is $-85\,\mathrm{kJ\,mol^{-1}}$. We could further complicate the problem by imagining an iron alloy in which case we'd need to know the standard free enthalpy of formation of the alloy, its iron content and the activity of Fe in the alloy.

9.3   Open-Circuit Voltage — Electromotive Force

Consider the electrochemical cell shown in Figure 9.1. On the left is a hydrogen electrode; the electrolyte is hydrochloric acid, HCl; the right-hand side electrode is silver on which silver chloride has been deposited in lumps by the action of the acid. The "cell diagram" is

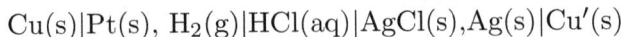

$$\mathrm{Cu(s)|Pt(s),\ H_2(g)|HCl(aq)|AgCl(s),Ag(s)|Cu'(s)}$$

The Pt and Ag electrodes are connected via copper wire through a high impedance voltmeter so that no current flows but the electric potential difference, $\phi_{\mathrm{Cu'}} - \phi_{\mathrm{Cu}}$, can be measured. We wish to calculate this "open-circuit" voltage and determine how much useful work can be done by the cell if a current is allowed to flow. By convention, the oxidation reaction is placed on the left and the reduction reaction on the right, and if current is allowed to flow, then positive charge will flow from left to right through the electrolyte and

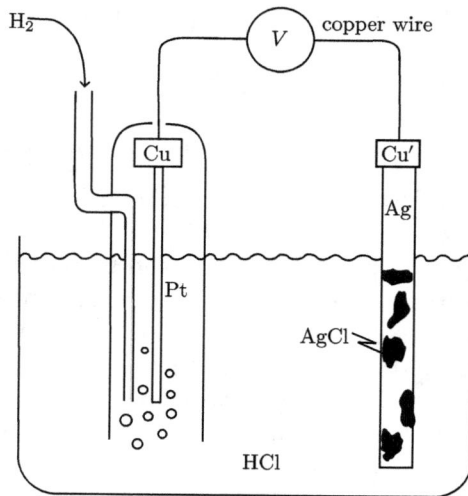

Fig. 9.1.   Silver/silver chloride electrode connected to a hydrogen electrode.

negative charge (electrons) will flow from left to right through the external circuit.

The oxidation reaction at the anode is

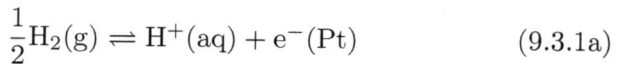

$$\frac{1}{2}H_2(g) \rightleftharpoons H^+(aq) + e^-(Pt) \tag{9.3.1a}$$

and the reduction reaction at the cathode is

$$AgCl(s) + e^-(Ag) \rightleftharpoons Ag(s) + Cl^-(aq) \tag{9.3.1b}$$

The combined reaction is

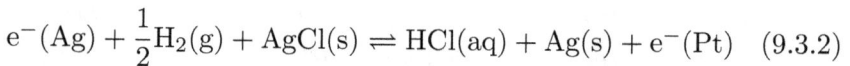

$$e^-(Ag) + \frac{1}{2}H_2(g) + AgCl(s) \rightleftharpoons HCl(aq) + Ag(s) + e^-(Pt) \tag{9.3.2}$$

The change in free enthalpy is

$$\Delta G = \mu_{HCl} + \mu_{Ag} + \tilde{\mu}_e^{Pt} - \left(\frac{1}{2}\mu_{H_2} + \mu_{AgCl} + \tilde{\mu}_e^{Ag}\right)$$

$$= \mu_{HCl}^{\square} + \mu_{Ag}^{\bullet} - \frac{1}{2}\mu_{H_2}^{\ominus} - \mu_{AgCl}^{\circ} + RT \ln \frac{h_{HCl}}{p^{1/2}} + \left(\tilde{\mu}_e^{Pt} - \tilde{\mu}_e^{Ag}\right) \tag{9.3.3a}$$

$$= \Delta G^{\circ} + RT \ln h_{HCl} + \left(\tilde{\mu}_e^{Pt} - \tilde{\mu}_e^{Ag}\right) \tag{9.3.3b}$$

in which h_{HCl} is the electrolyte activity on the practical Henrian scale of moles per kg of water (see Equations (A3.12a) and (A3.22a)) and p is the pressure of hydrogen as measured in bar. The molar free enthalpies of the pure hydrogen and silver can be taken as zero and the pressure of hydrogen is adjusted to one bar. AgCl is not a pure solid, but we assume for simplicity that it is in its standard state. In this way, we have the *standard free enthalpy* for the reaction (9.3.2):

$$\Delta G^{\circ} = \mu_{HCl}^{\square} - \mu_{AgCl}^{\circ}$$

When both the reduction and oxidation reactions (9.3.1) come to equilibrium, then at open circuit, the free enthalpy change is zero, $\Delta G = 0$, even though the electrons in the electrodes are not at equilibrium. However, since each electrode is in contact with its copper wire, the electrons across these junctions will have the same electrochemical potential and while

$$\tilde{\mu}_e^{Pt} = \tilde{\mu}_e^{Cu}; \quad \tilde{\mu}_e^{Ag} = \tilde{\mu}_e^{Cu'} \qquad (9.3.4)$$

the electrochemical potentials in the two wires connected across the voltmeter are not the same. At this stage, it's helpful to draw a ladder diagram as in Figure 2.3. This is shown in Figure 9.2. The electric

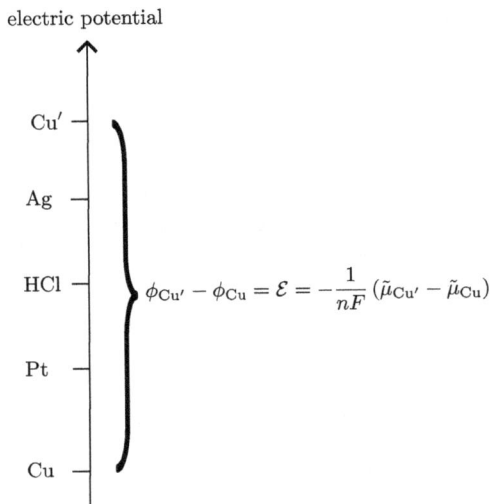

Fig. 9.2.   Ladder diagram for the cell of Figure 9.1.

potential of the anode is negative compared to the electric potential of the cathode. That is, $\phi_{Cu'} > \phi_{Cu}$. Conventionally represented as the electrode on the left, the anode is the electron sink (Figure 1.2), so a metal atom is losing an electron to become a cation which will migrate to the right through the electrolyte ("cations" migrate to the "cathode"). The electrons left behind being negatively charged serve to *lower* the electric potential on the anode.[4] The right-hand side electrode, the cathode, is conversely more positive than the left-hand side electrode. The total potential difference across the electrochemical cell, as indicated in Figure 9.2, is $\mathcal{E} = \phi_{Cu'} - \phi_{Cu} > 0$ and is called the "electromotive force", or electromotance, of the cell. It is measured in volts. The existence of this non-zero e.m.f. is due entirely to the difference in electrochemical potentials on the two electrodes. It must be true therefore that

$$\mathcal{E} = -\frac{1}{F}\left(\tilde{\mu}_e^{Ag} - \tilde{\mu}_e^{Pt}\right) = -\frac{1}{F}\left(\tilde{\mu}_e^{Cu'} - \tilde{\mu}_e^{Cu}\right)$$

Putting this fact into (9.3.3) with $\Delta G = 0$ at open circuit, and using (9.3.4), we arrive at the formula we are looking for, namely

$$-F\mathcal{E} = \Delta G^\circ + RT \ln h_{HCl} \qquad (9.3.5)$$

All the quantities in (9.3.5) can be measured: the concentration of HCl is ours to control and the standard molar free enthalpies can be looked up in tables of data to obtain the standard free enthalpy change, ΔG°, for the cell reaction.

As to the amount of work that the cell might do, we note that at open circuit, the cell reactions (9.3.1) are in equilibrium even though the electrochemical potentials of the electrons in the Pt and Ag electrodes are not. If we were to allow some current to flow through the external circuit, but *slowly*, then we may allow the cell to do some reversible work. If one mole of reactants is permitted to become one mole of products, then the change in free enthalpy is ΔG. If in the

---

[4]This is confusing because we learn that "anodes are positive and they corrode". True. If I turn the voltmeter into a battery and connect the positive terminal to the left-hand side electrode, this will remove electrons from the electrode, make its electric potential more positive and drive the oxidation (corrosion) of the electrode metal. In this case, the platinum is not corroding. The de-electronation reaction is the oxidation of hydrogen gas to hydrogen ions (protons).

process n moles of positive charge are transferred from left to right through the electrolyte, then an equal and opposite electronic charge, $-nF$, is transferred from left to right through the copper wire across an electric potential difference of $\Delta\phi = \mathcal{E}$, which will do an amount of work equal to $-nF\mathcal{E}$. *This statement lies at the heart of electrochemistry, batteries, fuel cells ...* This is exactly how *chemical energy* is converted into *electrical energy*, and of course the reverse is also true: if the charge on $-nF$ moles of electrons is carried from right to left by a power source, then one mole of product will be converted to one mole of reactant in the cell. This is how electricity is stored as chemical energy in a battery. Perhaps the most important equation in electrochemistry is the one that expresses the equivalence of chemical and electrical work — from the above argument, it is clear that

$$\Delta G = -nF\mathcal{E} \tag{9.3.6}$$

The meaning of the number, n, can be a bit confusing. It is the number of moles of electrons involved in the combined "redox" reaction. So in the case of (9.3.2), $n = 1$.

9.4 Single Electrode Potential

We've seen that we cannot measure a single electrode potential. Nevertheless, it is very useful to rank them by comparing different metal single electrodes to a reference electrode under standard conditions. The reference usually used is the standard hydrogen electrode (SHE). It is evident from Figures 2.3 and 9.2 that at least formally we can divide the electric potential drop across the cell into that due to the silver half cell and that due to the SHE

$$(\phi_{\text{Ag}} - \phi_{\text{Pt}}) = (\phi_{\text{Ag}} - \phi_{\text{HCl}}) + (\phi_{\text{HCl}} - \phi_{\text{Pt}})$$

or

$$^{\text{Ag}}\Delta^{\text{Pt}}\phi = {}^{\text{Ag}}\Delta^{\text{HCl}}\phi - {}^{\text{Pt}}\Delta^{\text{HCl}}\phi \tag{9.4.1}$$

Suppose I am interested in a particular metal, M, which may be oxidised in an electrolyte, S, containing M^{n+} metal ions having a charge of ne. I can construct an electrochemical cell just as shown in Figure 2.4, but with the SHE on the left,[5]

$$\text{Pt(s)}, \text{H}_2(\text{g}, p = 1\text{bar})|\text{H}^+(\text{aq, activity, } h_{\text{H}^+} = 1)||\text{M}^{n+}(\text{aq}, h_{\text{M}^{n+}})|\text{M(s)}$$

(the "||" is a salt bridge as in Figure 2.4) where the activities are practical "working" ones (see Equation (A3.26a), Section A.3.5). For this cell, I will have that the e.m.f. according to (9.4.1) is

$$\mathcal{E} = {}^{m}\Delta^{s}\phi - {}^{Pt}\Delta^{s}\phi = \mathcal{E}_{\text{M}}$$

essentially the difference between two half cell potentials. By definition, $\mathcal{E}_{\text{M}}$ is the *single potential* of the $\text{M}^{n+}(\text{aq})|\text{M(s)}$ electrode *on the SHE scale*. The cell reaction is

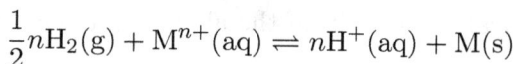

$$\frac{1}{2}n\text{H}_2(\text{g}) + \text{M}^{n+}(\text{aq}) \rightleftharpoons n\text{H}^+(\text{aq}) + \text{M(s)}$$

for which the free enthalpy change is

$$\Delta G = \mu_{\text{M}}^{\bullet} + n\mu_{\text{H}^+}^{\square} - \frac{1}{2}n\mu_{\text{H}_2}^{\bullet} - (\mu_{\text{M}^{n+}}^{\square} + RT\ln h_{\text{M}^{n+}})$$

$$= -(\mu_{\text{M}^{n+}}^{\square} + RT\ln h_{\text{M}^{n+}})$$

(pure substance standard chemical potentials being zero). Since we arbitrarily set the e.m.f. of the SHE to zero, we have, using (9.3.6),

$$\mathcal{E}_{\text{M}} = -\frac{\Delta G}{nF}$$

$$= \frac{\mu_{\text{M}^{n+}}^{\square}}{nF} + \frac{RT}{nF}\ln h_{\text{M}^{n+}} \tag{9.4.2a}$$

$$= \mathcal{E}_{\text{M}}^{\circ} + \frac{RT}{nF}\ln h_{\text{M}^{n+}} \tag{9.4.2b}$$

---

[5]The SHE is defined as having unit activity of hydrogen ions. This is problematic for me because even if the activity of an individual ion does exist in thermodynamics, it is certainly not measurable, see Section A.3.5. The best authority I can find is "Corrosion", edited by L. L. Shrier *et al.* (Butterworth Heinemann, 3rd edition 1994). There the SHE by definition has unit activity, on an unspecified scale, of H^+ ions which is identified as being the same as a molality of $m_{\text{H}^+} = 1.2$ moles HCl per kg water. See Section A.3.6. Moore (see Further Reading in Appendix D) gives the mean ion activity (A3.20) as one. Denbigh (see Further Reading in Appendix D) prefers "standard" to refer to unit concentration rather than activity; this makes it much easier in the laboratory to prepare a standard solution, but it emphasises that there is no universal agreement as to the amount of acid in the SHE.

Table 9.2.   Standard chemical potentials on the unimolal scale relative to H^+ ($kJ\,mol^{-1}$).

H^+	Ti^{++}	Ni^+	Al^{+++}	Fe^{++}	Fe^{+++}	Zn^{++}
0	−160	−50	−480	−85	−11	−150

Au^{+++}	Mg^{++}	Hg^+	Sn^{++}	Ag^+	Cu^+	Cu^{++}
+410	−450	+160	−25	+75	+50	+65

We have followed the usual convention and set $\mu^{\square}_{H^+} = 0$, which means that the standard unimolal chemical potential of the metal ion is implicitly referred to that of the hydrogen ion. Table 9.2 shows $\mu^{\square}$ for some cations in kilojoule per mole.

The more negative is the standard chemical potential, the more work the cation can do at the anode of an electrochemical cell.

9.5   Electrochemical or Electromotive Force Series

$\mathcal{E}^\circ_M$ is the *standard single-cell potential on the SHE scale*. These may be compiled into a ranked table called the *electrochemical series* or the *electromotive force series*. Table 9.3 is a compilation of standard single potentials. If at the first page of this book, the lemon lamp, you asked how is it that the zinc corrodes and not the copper; the answer is in the electrochemical series. When the two half cells $Zn^{++}(aq)|Zn(s)$ and $Cu^{++}(aq)|Cu(s)$ are connected electrically, and they share the same electrolyte (lemon juice which is acidic), then the $Zn^{++}(aq)|Zn(s)$ electrode having a more negative standard electrode potential (−0.76 V, compared to +0.34 V) becomes the anode and de-electronation occurs at the Zn electrode and electronation at the Cu electrode. As expressed in (9.4.2), $\mathcal{E}^\circ_{Zn} < \mathcal{E}^\circ_{Cu}$ because $\mu^{\square}_{Zn^{n+}} < \mu^{\square}_{Cu^{n+}}$, that is, the Zn ion has the more negative standard chemical potential in solution in water. Evidently, the system can gain more free enthalpy if the Zn dissolves, rather than the Cu. This to my mind does not really *explain* why the Zn chooses to corrode. You may gaze at the electrochemical series and wonder how nature has organised metals in this way. To a physicist, the noble metals are Cu, Ag and Au. To the chemist, the noble metals are those at the top of the electrochemical series: Au, Pt and Pd; they call Cu, Ag and Au the *coinage metals*. The metals with a large negative $\mathcal{E}^\circ_M$ are

Table 9.3.   The electrochemical or electromotive force series.

Redox couple	Standard e.m.f (volts)
$Au^{3+}(aq) + 3e^-$ (in Au) $\rightarrow$ Au(s)	1.50
$Pt^{2+}(aq) + 2e^-$ (in Pt) $\rightarrow$ Pt(s)	1.20
$Pd^{2+}(aq) + 2e^-$ (in Pd) $\rightarrow$ Pd(s)	0.99
$Hg^{2+}(aq) + 2e^-$ (in Hg) $\rightarrow$ Hg(s)	0.85
$Ag^+(aq) + e^-$ (in Ag) $\rightarrow$ Ag(s)	0.80
$Hg_2^{2+}(aq) + 2e^-$ (in Hg) $\rightarrow$ 2Hg(s)	0.79
$Cu^+(aq) + e^-$ (in Cu) $\rightarrow$ Cu(s)	0.52
$Cu^{2+}(aq) + 2e^-$ (in Cu) $\rightarrow$ Cu(s)	0.34
$Ge^{2+}(aq) + 2e^-$ (in Ge) $\rightarrow$ Ge(s)	0.23
$2H^+(aq) + e^-$ (in Pt) $\rightarrow$ H_2(gas, 1bar)	0.00
$Pb^{2+}(aq) + 2e^-$ (in Pb) $\rightarrow$ Pb(s)	−0.13
$Sn^{2+}(aq) + 2e^-$ (in Sn) $\rightarrow$ Sn(s)	−0.14
$Ni^{2+}(aq) + 2e^-$ (in Ni) $\rightarrow$ Ni(s)	−0.25
$Co^{2+}(aq) + 2e^-$ (in Co) $\rightarrow$ Co(s)	−0.28
$Tl^+(aq) + e^-$ (in Tl) $\rightarrow$ Tl(s)	−0.34
$In^{3+}(aq) + 3e^-$ (in In) $\rightarrow$ In(s)	−0.34
$Cd^{2+}(aq) + 2e^-$ (in Cd) $\rightarrow$ Cd(s)	−0.40
$Fe^{2+}(aq) + 2e^-$ (in Fe) $\rightarrow$ Fe(s)	−0.44
$Ga^{3+}(aq) + 3e^-$ (in Ga) $\rightarrow$ Ga(s)	−0.53
$Cr^{3+}(aq) + 3e^-$ (in Cr) $\rightarrow$ Cr(s)	−0.74
$Cr^{2+}(aq) + 2e^-$ (in Cr) $\rightarrow$ Cr(s)	−0.91
$Zn^{2+}(aq) + 2e^-$ (in Zn) $\rightarrow$ Zn(s)	−0.76
$V^{2+}(aq) + 2e^-$ (in V) $\rightarrow$ V(s)	−1.18
$Mn^{2+}(aq) + 2e^-$ (in Mn) $\rightarrow$ Mn(s)	−1.18
$Zr^{4+}(aq) + 4e^-$ (in Zr) $\rightarrow$ Zr(s)	−1.53
$Ti^{2+}(aq) + 2e^-$ (in Ti) $\rightarrow$ Ti(s)	−1.63
$Al^{3+}(aq) + 3e^-$ (in Al) $\rightarrow$ Al(s)	−1.66
$Hf^{4+}(aq) + 4e^-$ (in Hf) $\rightarrow$ Hf(s)	−1.70
$U^{3+}(aq) + 3e^-$ (in U) $\rightarrow$ U(s)	−1.80
$Be^{2+}(aq) + 2e^-$ (in Be) $\rightarrow$ Be(s)	−1.85
$Mg^{2+}(aq) + 2e^-$ (in Mg) $\rightarrow$ Mg(s)	−2.37
$Na^+(aq) + e^-$ (in Na) $\rightarrow$ Na(s)	−2.71
$Ca^{2+}(aq) + 2e^-$ (in Ca) $\rightarrow$ Ca(s)	−2.87
$K^+(aq) + e^-$ (in K) $\rightarrow$ K(s)	−2.93
$Li^+(aq) + e^-$ (in Li) $\rightarrow$ Li(s)	−3.05

called *base metals*. To a corrosion scientist, the fact that a metal of interest, for example, magnesium, is a base metal can be depressing. In the design of Mg alloys, poor corrosion resistance is one of the principal reasons why this least dense of engineering alloys is less ubiquitous than you may expect. But there is almost nothing that can be done to combat the fundamental fact that $\mathcal{E}^\circ_{Mg} = -2.37\,\text{V}$. I have no deep insight as to why the metals are ordered as they are. Fawcett, in his textbook (see Further Reading in Appendix D), gives three factors that determine the position of a metal in the series: (i) the free enthalpy of solvation of the metal ion in water; (ii) the free enthalpy of the metal, which depends on its crystal and electronic structure; (iii) the ionisation energy of the metal atom. I'll return to this in Section 14.6.

9.6 Nernst Equation

The half cell reaction can be generalised to consider a substance "Ox", which is reduced by acquiring an electron from the electrode to become the reduced substance, "Red":

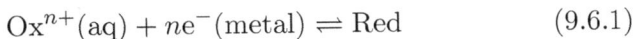

$$Ox^{n+}(aq) + ne^-(metal) \rightleftharpoons Red \tag{9.6.1}$$

The half cell single potential can be measured as in Section 9.4 by connecting to a SHE:

$$Pt(s), H_2(g,\ p = 1\text{bar})|H^+(aq,\ \text{activity},\ a = 1)||Ox^{n+}(aq)|Red, \text{metal}$$

The cell reaction is

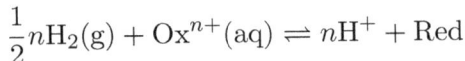

$$\frac{1}{2}nH_2(g) + Ox^{n+}(aq) \rightleftharpoons nH^+ + Red$$

The e.m.f. is

$$\mathcal{E} = \mathcal{E}^\circ + \frac{RT}{nF} \ln \frac{a_{Ox}}{a_{Red}} \tag{9.6.2}$$

The standard e.m.f. is

$$\mathcal{E}^\circ = \frac{1}{nF}\left(\mu^\circ_{Ox} - \mu^\circ_{Red}\right) \tag{9.6.3}$$

Equation (9.6.2) is called the *Nernst equation*. The numerical value of the standard e.m.f. will depend on the scale and convention that are used for the activity (see Section A.3). We return to the Nernst equation in Section 14.7.

Chapter 10

Relative and Absolute Ion Work Function and Real Potential

10.1 Measurement of the Proton Work Function and Real Potential in Water

The electrochemical potential of a charged species is *in principle* not measurable. However, measurements of its real potential can be made if assumptions are made that extend beyond the scope of equilibrium thermodynamics. Many measurements are made using a device called "Kenrick's apparatus." The idea is to create an air gap between a metal electrode and an electrolyte and create conditions in which the Volta potential difference between the two surfaces is zero. This can be done as already described in Section 3.4 using the Kelvin probe. There, a bias potential is applied of an amount just to prevent any current flowing if the surfaces are moved closer to each other or further away. That so called *compensation potential,* denoted $\Delta_c\phi$, is equal to minus the difference in Volta potential, permitting the measurement of the Volta potential difference between two surfaces. In Kenrick's apparatus, rather than applying a bias potential, it is arranged that the Volta potential difference is zero by design. This is achieved using a liquid mercury electrode that is arranged to flow rapidly past a stream of electrolyte flowing in the same direction, with an air gap in between. Because the two opposing surfaces are in rapid flow, it is assumed that charge cannot distribute in time so that the Volta potential difference across the air gap is zero. I have to say I do not find this convincing: the time taken for charge to distribute in a metal must be roughly the inverse plasma frequency; I don't know what it is in an electrolyte. But if the system is in steady state, or worse, then I don't see how we can apply thermodynamic principles. However, it's been around a long time and it works. (You really can abuse thermodynamics if you don't mind offending the ghost of Josiah Willard Gibbs — when I was a metallurgy undergraduate doing problems, they occasionally required the student to "assume equilibrium in the blast furnace".) Using this apparatus, and making one very reasonable and not at all serious non-thermodynamic assumption, it is possible to measure the *work function of the proton in water*. The assumption is necessary because the activity of a proton cannot be measured. So in hydrochloric acid, HCl, it is assumed that the activity of the H^+ ion is equal to the mean ion activity. The Kenrick apparatus for this experiment uses streams of mercury and HCl, the electrolyte being connected to a standard hydrogen electrode. This is illustrated in Figure 10.1. The cell diagram is

$$\text{Cu} | \text{Hg} | \text{air} | \text{HCl, activity } a_{HCl} | \text{Pt, } H_2 \ p = 1 \text{ bar} | \text{Cu}'$$

The potential measured across the copper leads is a compensation potential by virtue of

$$\psi_s - \psi_{Hg} = 0$$

and is given by

$$
\begin{aligned}
F\Delta_c\phi &= \tilde{\mu}_e^{Cu} - \tilde{\mu}_e^{Cu'} \\
&= \tilde{\mu}_e^{Hg} - \tilde{\mu}_e^{Pt} \\
&= \mu_e^{Hg} - \mu_e^{Pt} - F(\phi_{Pt} - \phi_{Hg})
\end{aligned}
\tag{10.1.1}
$$

By considering the equilibrium of the SHE half cell, using procedures just as in Chapter 9, one finds for $^{HCl}\Delta^{Pt}\phi$,

$$F(\phi_s - \phi_{Pt}) = \frac{1}{2}\mu_{H_2}^{g,o} - \mu_{H^+}^{s,o} - \mu_e^{Pt} - RT\ln a_{H^+}$$

where a_{H^+} is the activity of the hydrogen ion (we are not specifying any particular reference standard state and are using the

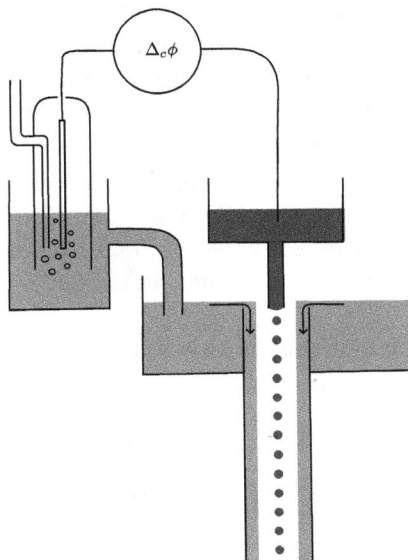

Fig. 10.1. Kenrick's apparatus. At the top left is a hydrogen electrode feeding hydrochloric acid at a potential ϕ_s into the vessel at the right. Above the vessel is a reservoir of mercury which is at potential ϕ_{Hg}. It's all going on in the exit tube: HCl is pouring down the sides of the tube, while drops of Hg are falling down the centre. The flow assures that the outer or Volta potential difference across the air gap is zero, and therefore what is measured at the voltmeter is a *compensation potential*, $\Delta_c\phi$.
Source: Adapted with permission from Ronald Fawcett, *Langmuir*, **24**, 9868 (2008).

generic ° superscript). Combining this with (10.1.1) to eliminate Pt potentials in favour of Hg potentials results in

$$F\Delta_c\phi = F\left(\chi_s - \chi_{Hg}\right) + RT\ln a_{H+} - \Delta G^\circ$$

where

$$\Delta G^\circ = \frac{1}{2}\mu_{H_2}^{g,\circ} - \mu_{H+}^{s,\circ} - \mu_e^{Hg}$$

Here the dipole potential difference enters since in view of the air gap and the flowing electrode and electrolyte, the Volta potential difference is zero and ${}^s\Delta^{Hg}\phi = {}^s\Delta^{Hg}\chi$. The experimenter measures the voltage across the copper leads with a very high impedance voltmeter as a function of the strength of the acid. Assuming that the activity of H^+ is equal to the mean ion activity which can be measured (this is the "extrathermodynamic" assumption: $a_{H^+} = a_\pm = \sqrt{a_{HCl}}$, Section A.3.5, Equation (A3.24a)), then plotting $\Delta_c\phi - RT\ln m_\pm$, where $m_\pm = m_{HCl}$ is the molal concentration of the acid (assuming in the dilute limit that the mean ion activity coefficient is one) against the square root of the HCl concentration and extrapolating to zero concentration yields a value of $55.9 \pm 0.2\,\text{mV}$ in the infinitely dilute limit. Because it's constant, we don't need to know what $\Delta G°$ is.

Next, employ the fact that in this limit the dipole potential of the infinitely dilute acid is the same as that of water, χ_{H_2O}. Then,

$$F\Delta_c\phi - RT\ln m_\pm \, (m_\pm \to 0) = -55.9\,\text{mV} \times F$$

$$= \mu_{H^+}^{s,°} + F\chi_{H_2O} + \mu_e^{Hg} - F\chi_{Hg} - \frac{1}{2}\mu_{H_2}^{g,°}$$

$$= \alpha_{H^+}^{s,°} + \alpha_{e,Hg}^{s,°} - \frac{1}{2}\mu_{H_2}^{g,°} \qquad (10.1.2)$$

using, in the third line, the definition of the real potential (4.2.1) and remembering that the charge on the electron is negative. Now, the real potential of the electron in a metal is the same as minus its work function, which for mercury is $4.5\,\text{eV}$. The last term in (10.1.2) is the standard chemical potential of hydrogen in the gas phase which is zero for the pure substance. Hence (see Footnote 3 of Section 9.2),

$$\alpha_{H^+}^{s,°} = -0.056 + 4.5 = 4.44\,\text{eV} \qquad (10.1.3)$$

The proton work function is (8.2.8)

$$W_{H^+}^{s,°} = \mu_{H^+}^{g,°} - \alpha_{H^+}^{s,°}$$

The first term, $\mu_{H^+}^{g,°}$, is the standard free enthalpy of formation of a proton in the gas phase. Relative to the standard free enthalpy of the hydrogen molecule, this is half the dissociation (bond) energy $(4.52\,\text{eV})$ plus the ionisation energy $(13.6\,\text{eV})$, namely $15.9\,\text{eV}$. The work function of the proton in water then comes out as $11.43\,\text{eV}$ $(1103\,\text{kJ/mol})$. Once that is known, then using (8.2.6) and the measured standard free enthalpy of solvation of hydrogen ions (again not strictly measurable but using sensible extrathermodynamic assumptions), it is possible to deduce from experiment that the electric dipole potential at the surface of water is 20 mV (see Problem 10.1).

10.2 Absolute Electrode Potential

The electrochemical series ranks electrode potentials and provides a single electrode potential *relative to the standard hydrogen single electrode potential*. This is inevitable because thermodynamics does not furnish us with an absolute zero of energy. Hence, we choose the one most convenient to us. So far we have encountered at least five choices. (*i*) In the DFT using periodic boundary conditions and hence with no reference to the surfaces of the substance or the world beyond it, we use the average electric potential, $-\overline{V}/e$, seen by an electron to define $\overline{V}$ as a convenient reference (Section 3.1). (*ii*) The DFT total energy, E_{tot}, is calculated with reference to the situation where all the electrons and nuclei are scattered, each to its own field-free infinity (the nuclei are not disassembled into nucleons or quarks however). (*iii*) We have used the electric potential at infinity or at the vacuum level ("Bockris point") as convenient zeros of potential. (*iv*) The reader may be bemused and sceptical of our conveniently setting certain standard chemical potentials to zero. We can do this if we allow that the zero of energy is somewhat similar to the DFT zero, but instead we envisage a number of reservoirs of pure substance, solid or ideal gas at one bar and 298 K, infinitely far from each other and from which we are drawing our components to make up the substance under study. (*v*) We have used the single SHE potential as a zero of electric potential. Electrochemists also use other standards such as the standard calomel electrode, but they can easily be calibrated to the SHE by constructing a suitable electrochemical cell.

The question now is, can we find an *absolute* single electrode potential for the SHE relative to the thermodynamic reference (*iv*)? This cannot be done without using assumptions beyond the equilibrium thermodynamics, but in fact there is an argument first put forward by Sergio Trasatti in the 1980s.

We start by looking at Section 9.3 in a different way. There we used the fact that the e.m.f. of the cell in Figure 9.1 was the difference in electrochemical potentials of the electrons in the right and left hand leads. Instead, pursue the following line of argument. The equilibrium in the left-hand side cell (9.3.1a) implies equality of the electrochemical potentials:

$$\frac{1}{2}\mu_{H_2} = \tilde{\mu}_{H^+} + \tilde{\mu}_e^{Pt}$$

$$= \mu_{H^+} + F\phi_s + \mu_e^{Pt} - F\phi_{Pt}$$

$$= \mu_{H^+}^{\square} + RT\ln h_{H^+} + F\phi_s + \mu_e^{Pt} - F\phi_{Pt}$$

This is the *standard* hydrogen electrode, so we'll have unit activity of H^+ and 1 bar pressure of hydrogen gas. Then after rearranging,

$$F(\phi_{Pt} - \phi_s) = -\frac{1}{2}\mu_{H_2}^{\ominus} + \mu_{H^+}^{\square} + \mu_e^{Pt} \qquad (10.2.1)$$

Equilibrium of the right-hand side cell (9.3.1b) similarly implies the equality of electrochemical potentials on either side of the reaction,

$$\mu_{AgCl} + \tilde{\mu}_e^{Ag} = \mu_{Ag} + \tilde{\mu}_{Cl^-}$$

and after expanding the electrochemical potentials and rearranging exactly as above,

$$F(\phi_s - \phi_{Ag}) = \mu_{Ag}^{\bullet} - \mu_{AgCl}^{\bullet} + \mu_{Cl^-}^{\square} + RT\ln h_{Cl^-} - \mu_e^{Ag}$$

$$= -\mu_{AgCl}^{\bullet} + \mu_{Cl^-}^{\square} + RT\ln h_{Cl^-} - \mu_e^{Ag} \qquad (10.2.2)$$

since Ag is in its pure substance standard state. The open-circuit e.m.f. is

$$\mathcal{E} = \phi_{Cu'} - \phi_{Cu}$$

$$= (\phi_{Pt} - \phi_{Cu}) + (\phi_s - \phi_{Pt}) + (\phi_{Ag} - \phi_s) + (\phi_{Cu'} - \phi_{Ag}) \qquad (10.2.3)$$

In view of (3.2.1) and Footnote 3 of Section 9.2, we have

$$\phi_{Pt} - \phi_{Cu} = \frac{1}{F}\left(\mu_e^{Pt} - \mu_e^{Cu}\right)$$

$$\phi_{Cu'} - \phi_{Ag} = \frac{1}{F}\left(\mu_e^{Cu'} - \mu_e^{Ag}\right)$$

A key point is that the Fermi levels in both copper leads are the same even though their electrons are at different electrochemical potentials. This is because the Fermi level is a property of the bulk crystal (Figure 3.1). So $\mu_e^{Cu'} = \mu_e^{Cu}$ and (10.2.3) becomes

$$\mathcal{E} = \left((\phi_{Ag} - \phi_s) - \frac{1}{F}\mu_e^{Ag} \right) - \left((\phi_{Pt} - \phi_s) - \frac{1}{F}\mu_e^{Pt} \right)$$

$$= \mathcal{E}_{Ag|AgCl} - \mathcal{E}_{H^+|Pt} \qquad (10.2.4)$$

The following two points are worth noting.

(i) Comparing (10.2.4) with (10.2.1) and (10.2.2), it is clear that the Fermi levels in each metal cancel. This will always happen in a formula for the full cell e.m.f. but not in the formula for a half cell, such as (9.2.1). However, because of this cancellation, you will see in textbooks such as John West's a statement like, "we assume that electrons in metals are in their standard state so their standard chemical potentials can be set to zero."

(ii) In spite of this, we leave (10.2.4) in the form that it is because it has separated the e.m.f. into single electrode potentials, each depending only on the properties of its own electrode.

Because of point (ii), we could imagine adding and subtracting a constant, K_{abs}, to (10.2.4) as an attempt to identify an *absolute* electrode potential:

$$\mathcal{E}_{Ag|AgCl}^{abs} = (\phi_{Ag} - \phi_s) - \frac{1}{F}\mu_e^{Ag} + K_{abs}$$

$$\mathcal{E}_{H^+|Pt}^{abs} = \mathcal{E}_{SHE}^{abs} = (\phi_{Pt} - \phi_s) - \frac{1}{F}\mu_e^{Pt} + K_{abs}$$

Now, Trasatti's idea is to modify the cell in Section 9.3 so as to introduce an air gap into the electrolyte:

$$Cu(s)|Pt(s), H_2(g)|HCl(aq)|air|HCl(aq)|AgCl(s), Ag(s)|Cu'(s)$$

This is a thought experiment so maybe it doesn't matter how to realise this in practice, but it is imagined that the Volta potential difference across the air gap can be made to be zero, presumably by some streaming of the electrolyte as in Kenrick's apparatus (Section 10.1). In that case, the e.m.f. at open circuit is a compensation potential (Section 10.1) and in the manner of (10.2.3), this

can be written as a sum of potential differences across the phase
boundaries indicated by "|" in the cell diagram:

$$\Delta_c\phi = (\phi_{Cu'} - \phi_{Cu})$$
$$= (\phi_{Pt} - \phi_{Cu}) + (\phi_s - \phi_{Pt}) + (\phi_{air} - \phi_s)$$
$$+ (\phi_s - \phi_{air}) + (\phi_{Ag} - \phi_s) + (\phi_{Cu'} - \phi_{Ag})$$

Because of the zero Volta potential difference across the air gap, we
have

$$(\psi_{air} - \psi_s) = 0$$

and so the Galvani potential difference is just the dipole potential
difference between the HCl electrolyte and air (we know that the
surface dipole potential of water is about 20 mV, Section 10.1, but
we don't need to use this number in what follows):

$$(\phi_{air} - \phi_s) = \chi_s$$

Going through the same steps leading to Equations (10.2.1), (10.2.2)
and (10.2.4), we arrive at

$$\mathcal{E} = \left((\phi_{Ag} - \phi_s) - \frac{1}{F}\mu_e^{Ag} + \chi_s\right) - \left((\phi_{Pt} - \phi_s) - \frac{1}{F}\mu_e^{Pt} + \chi_s\right)$$
$$= \mathcal{E}_{Ag|AgCl}^{abs} - \mathcal{E}_{H+|Pt}^{abs}$$

We can assert that we have now found the absolute single SHE potential because the air gap and the high impedance across the voltmeter
have isolated the two half cells. In this case, using $K_{abs} = \chi_s$ identifies the constant needed to refer the SHE to the "thermodynamic"
zero of energy. Now using (10.2.1), we get

$$\mathcal{E}_{SHE}^{abs} = \phi_{Pt} - \phi_s - \frac{1}{F}\mu_e^{Pt} + \chi_s$$
$$= \frac{1}{F}\left(\mu_{H+}^\square - \frac{1}{2}\mu_{H_2}^\ominus + \mu_e^{Pt}\right) - \frac{1}{F}\mu_e^{Pt} + \chi_s$$
$$= \frac{1}{F}\mu_{H+}^\square + \chi_s$$
$$= \frac{1}{F}\alpha_{H+}^{s,\circ}$$

In the second line, the Fermi energy of Pt has cancelled as promised.
We also take the chemical potential of hydrogen gas in its standard

state of one bar pressure and 298 K as zero. The last line follows from (4.2.1) and leaves us with the remarkably simple result that the absolute e.m.f. of the standard hydrogen electrode is the real potential of the proton in the electrolyte. Assuming that this is the same as for water, $\chi_s = \chi_{H_2O}$ (and I don't see how we can assume this — 1.2 molal HCl is not dilute), then we have already calculated the answer in (10.1.3):

$$\mathcal{E}_{SHE}^{abs} = 4.44\,\text{V} \qquad (10.2.5)$$

This should be a great help in computer simulations of half cells, when comparing to measurements against the SHE. An alternative derivation and a refinement of the answer to 4.42 V can be found in Cheng and Sprik (see Further Reading in Appendix D).

Chapter 11

Electrode Capacitance and Electrocapillarity

One of the most striking observations in electrochemistry is the parallel between the electrical behaviour of the electrode–electrolyte interphase and an equivalent electrical circuit. We are reminded of the mechanical behaviour of polymers which can be mapped onto "circuits" made up of combinations of Maxwell and Voigt elements in series and parallel. The equivalent electrical circuits in electrochemistry are usually combinations of capacitors, resistors and mass transfer impedances, rather similar to the analogy between a mechanical damped oscillator and an LCR electrical circuit. To understand this, we need to look in detail into the structure and thermodynamics of the interphase.

11.1 Thermodynamics of the Interphase

Interfacial thermodynamics is a rather difficult subject. In fact, it has led some authors and even textbooks into confusion and indeed error. Here I follow John Cahn (see Further Reading in Appendix D) and Mike Finnis (see Further Reading in Appendix D). The subject was of course rigorously addressed by Josiah Willard Gibbs and it is not often that you come across a work that improves on Gibbs. But Cahn is a case in point and his theory is both elegant and compact while at the same time exposing neatly where error has arisen.

In order to avoid curvature terms in the free enthalpy, we will consider only a flat, planar interface between two phases α and β. These may be two electrolytes or an electrolyte and a metal electrode. A key point is that when the two phases are connected to create an interface or "interphase", at equilibrium the chemical potentials of each component throughout the assembly must remain the same as in the bulk separated phases; hence the mole numbers of each component as well as the other extensive quantities, in particular volume and entropy, will need to adjust. This gives rise to the notion of *excess* extensive properties compared to the amounts in the unconnected phases.[1]

Figure 11.1(a) is a cartoon representing a part of the interface between two phases, α and β. We choose as our *system*[2] the piece of matter which is a cylinder whose axis is orientated perpendicular to the interface separating the two phases. The cross-sectional area of the cylinder is A (the system doesn't have to have a circular cross-section: any prism will do). As I will show in a moment, the length of the cylinder is arbitrary as long as its ends lie in homogeneous phases, α and β, that is to say, the matter at the ends has all the properties of either bulk phase and is not affected by the presence of the interface. We follow Cahn and Finnis and treat this as a *closed* system (see Section A.2.2). We extend Equation (A2.5) as follows. If the interface has an area A, the combined first and second laws read

$$\mathrm{d}U = T\mathrm{d}S - p\mathrm{d}V + \mu_i \mathrm{d}n_i + \gamma\,\mathrm{d}A \qquad (11.1.1)$$

The third term contains the chemical potentials, μ_i (Equation (A2.4)), and n_i is the total number of moles of component (species) i in the system. The sum over i will be left implicit as in the Einstein summation convention. In the fourth term the interfacial energy, or surface tension, γ, appears in a work term: if the area is increased by $\mathrm{d}A$, then an amount of work $\gamma\,\mathrm{d}A$ is done on

---

[1]In thermodynamics, there are two meanings of "excess". One is as an excess of some quantity compared to a fictitious system which is ideal and is not universally regarded as at all useful. The second interpretation is the one we use here and the "excess over the ideal" is not a concept I will use.

[2]Recall Bockris's words: "All thermodynamic thinking begins with a definition of the portion of the universe under study, i.e., the system."

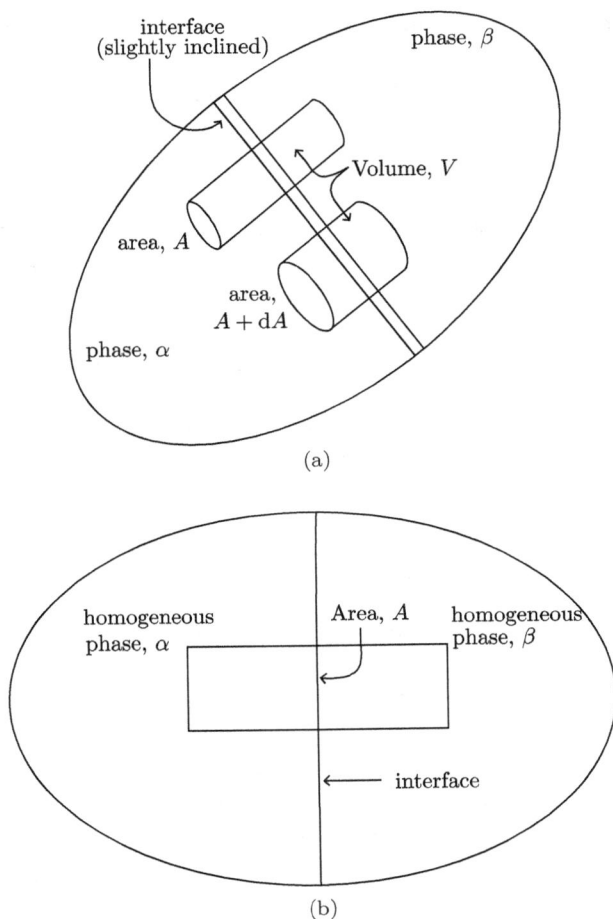

Fig. 11.1. (a) A piece of a body including an interface between two phases is enclosed by the outer line. The *system* we choose is a cylinder of matter enclosing a part of the interface orientated so that the interface is perpendicular to the axis of the cylinder. A way to increase the area at constant volume, while insisting the system is *closed*, is to redefine the system by varying its aspect ratio at constant volume. (b) Taking a sight along a line lying in the interface, we can draw our system in projection as a rectangle containing the interface parallel to the end sides. These ends are sufficiently far from the interface so as to exist in phases α and β having the same properties as the two isolated, homogeneous phases before they were brought together to form an interphase.

the system. It is vital to appreciate what is meant by increasing the area. This may not be done by stretching; new surface must be created. For example, think of two immiscible liquids in an upright test tube. If the tube is tilted, then the interfacial area increases, and although you don't realise it, in tilting the tube you have performed the work needed to increase the interfacial area. This is impossible if one or both of the phases are solid. This does introduce a difficulty: we wish to treat the system as closed, but we can't think of even a thought experiment in which the interfacial area is increased at constant volume. Finnis circumvents this issue by allowing that a perfectly legitimate process is simply to redraw the boundary of the system in Figure 11.1(a) by making it narrower and taller so as to keep the total volume fixed but increasing the area, A.

Equation (11.1.1) can be integrated to give

$$U = TS - pV + \mu_i n_i + \gamma A$$

so that the interfacial tension is seen to be

$$\gamma = \frac{1}{A}\left(U - TS + pV - \mu_i n_i\right) = \frac{1}{A}\left(G - \mu_i n_i\right) \qquad (11.1.2)$$

This has a simple interpretation. G is the free enthalpy of the system with interphase enclosed by the rectangular boundary in Figure 11.1(b) and $\mu_i n_i$ is the free enthalpy of the separated parts before assembly in view of Equation (A2.10) and in view of the point that before and after assembly, *in equilibrium* the chemical potentials are those of the components in the bulk homogeneous phases. This exposes the interfacial tension as an *excess interfacial free enthalpy.*

Although we have allowed that the choice of boundary, shown as a rectangle in Figure 11.1(b), is arbitrary, for all that follows it is necessary to insist that the horizontal length is such that the system encloses enough material to either side so that material intersected by the left- and right-hand side boundaries has exactly the properties of the homogeneous phases α and β before they were joined to form the interface. Under those restrictions only, the horizontal length of the system — the rectangle in Figure 11.1(b) — is arbitrary. You can see this from Equation (11.1.2). Suppose I decide to extend the left-hand side boundary by adding more moles of homogeneous phase α to the system. *This doesn't change the interfacial tension* because an equal amount, say $\mu_i \Delta n_i^\alpha$, is added to the two terms in (11.1.2), G and $\mu_i n_i$, which cancel out. An equivalent way of seeing this is to ask the following question: what if I have a surface and I add a further layer of atoms? This doesn't change the surface energy because the atoms that *were* on the surface are *now* subsurface and nothing has changed as regards the properties of the surface. A mathematical way to see this is that since in the homogeneous phases we must have $G = n_i \mu_i$, we must have in each phase that

$$U^\alpha - TS^\alpha + pV^\alpha - n_i^\alpha \mu_i = 0$$
$$U^\beta - TS^\beta + pV^\beta - n_i^\beta \mu_i = 0$$

In Appendix A, we defined molar extensive quantities by using a lower case symbol, for example, v is the molar volume. We now similarly define what Cahn calls *layer quantities* by dividing the internal energy, the volume, the entropy and the number of moles of component i in the system by the area A. I will deviate from Cahn's and Finnis's notation and denote these u, v, s and n. Individually, these are meaningless because of the arbitrary choice of width of the layer (the separation of the vertical lines in Figure 11.1(b)); however as long as they appear in certain *proper* combinations, they can contribute to measurable thermodynamic excess quantities. For example, the interfacial tension, using (11.1.2), is

$$\gamma = u - Ts + pv - n_i \mu_i$$

from which, by differentiating and combining with (11.1.1) divided through by A, can be obtained a version of the "Gibbs adsorption equation":

$$d\gamma = -sdT + vdp - n_i d\mu_i \qquad (11.1.3a)$$

Because the layer quantities appear in a *proper* combination, this is independent of the size of the layer (interphase) on account of the argument just given which carries through from (11.1.2).

Cahn points out how Equation (11.1.3a) can lead one into error. The reader may be tempted to assert that at constant T and p

(11.1.3a) becomes the Gibbs adsorption isotherm,[3]

$$\left(\frac{\partial \gamma}{\partial \mu_i}\right)_{T,p,\mu_j(j\neq i)} = -\mathsf{n}_i \quad \text{(wrong)} \tag{11.1.3b}$$

but this contains two flaws. Firstly, the right-hand side is a meaningless layer quantity. Secondly, the variation at constant temperature, pressure and chemical potential of all but one species cannot be made without violation of the Gibbs phase rule. We will see below how to write a correct adsorption isotherm. At the heart of this is that any variation must be made by reference to the Gibbs–Duhem equations (A2.11a) — one for each phase:

$$-S^\alpha dT + V^\alpha dp - n_i^\alpha \, d\mu_i = 0$$
$$-S^\beta dT + V^\beta dp - n_i^\beta \, d\mu_i = 0 \tag{11.1.4}$$

Each Gibbs–Duhem equation states that for each phase in equilibrium, independent variations of the intensive properties are not possible. You might think that if a system comprises C components, each having its chemical potential, and there are two additional intensive variables, T and p, then there will be $C+2$ degrees of freedom (number of intensive variables that can be adjusted independently while remaining in equilibrium). However, for each phase there is a Gibbs–Duhem equation which removes a degree of freedom, so the Gibbs phase rule (discovered by Gibbs between 1875 and 1878) states that the number of degrees of freedom is actually

$$F = C + 2 - P$$

where P is the number of phases (number of constraints). For example, a binary system consisting of two components and four phases (say, two solid alloy compositions and a liquid and gas phase all in equilibrium) can only exist at a single temperature and pressure and at fixed values of the two chemical potentials. Pure water has one component, so at the triple point where liquid, solid and vapour are in equilibrium, there are $F = 1 + 2 - 3 = 0$ degrees of freedom and

---

[3]You might cheekily look in some textbooks (even in Further Reading) to see if you can find this error.

so the temperature and pressure at the triple point are fixed, as is well known. Along the coexistence line separating liquid and vapour between the triple point and the critical point, there are two phases and so just *one* of either T or p can be adjusted and the other is then determined by the requirement to remain on the coexistence line. These are all examples of the phase rule at work.

11.1.1 *The case of a single phase*

Before embarking on the full machinery of Cahn's reformulation of Gibbs's interfacial thermodynamics, we follow Finnis and consider the special case of a single phase. This covers a number of simple cases: (i) a grain boundary, twin, stacking fault, anti-phase boundary where there is the same crystal on either side of the interface, and components may segregate to the boundary in order to maintain the same chemical potential as in the homogeneous phases, resulting in an *excess* interfacial concentration; (ii) the surface separating phase α with vacuum; (iii) the electrolyte–metal interphase. The last example is perhaps surprising but as long as the components of interest (ions and water) do not dissolve in the metal, then the quantities n_i^β are zero if we denote the metal phase by β so it effectively acts like vacuum. Therefore, this simplification is both revealing and relevant.

We have for the phase α the Gibbs adsorption equation and the Gibbs–Duhem equation:

$$\mathrm{d}\gamma = -\mathrm{s}\mathrm{d}T + \mathrm{v}\mathrm{d}p - \mathsf{n}_i\mathrm{d}\mu_i$$
$$0 = -S^\alpha \mathrm{d}T + V^\alpha \mathrm{d}p - n_i^\alpha \mathrm{d}\mu_i \qquad (11.1.5)$$

I cannot suppress the superscript α even though there is only one phase because I must distinguish between S, V and n_i which are the total entropy, volume and number of moles of component i of the system enclosed by the boundary in Figure 11.1(b) and S^α, V^α and n_i^α which are the entropy, volume and number of moles of component i of the homogeneous phase α, before bringing the two phases together to create the interface. Now, multiply the second Equation (11.1.5) by n_1/n_1^α in which we might choose for component 1 the solvent, or majority species, if there is one. Then subtract the result from the first equation which evolves into

$$d\gamma = -\left(s - \frac{n_1}{n_1^\alpha}S^\alpha\right)dT + \left(v - \frac{n_1}{n_1^\alpha}V^\alpha\right)dp - \left(n_i - \frac{n_1}{n_1^\alpha}n_i^\alpha\right)d\mu_i$$

(11.1.6)

Note that the implicit sum now effectively excludes $i = 1$ because that term is zero so the constraint has removed our freedom to vary the chemical potential of the (solvent) component 1. Now we are safe and the $C + 1$ variations of intensive quantities can be made independently; the chemical potential of component 1 will have to adjust itself in order to be consistent with the Gibbs–Duhem relation. The quantities in parentheses are known as the *interfacial excesses with respect to component 1*. In particular, the molar excess of component i with respect to component 1 is

$$\Gamma_i = n_i - \frac{n_1}{n_1^\alpha}n_i^\alpha$$

(11.1.7a)

The solvent excess is by construction zero.

For a solution in contact with gas phase,[4] the Gibbs adsorption isotherm follows from (11.1.6), namely

$$\left(\frac{\partial\gamma}{\partial\mu_i}\right)_{T,p,\mu_j(j>1,j\neq i)} = -\Gamma_i \quad (i > 1)$$

(11.1.8)

This is a correct variation of γ which can be made without violating the phase rule, and the right-hand side is independent of the size of the layer. An example of the Gibbs adsorption isotherm at work is the effect of soap on water. If a small amount of soap, component i, is dissolved in water, then if it segregates to the surface, its excess, Γ_i, will be positive. If I now add more soap, then the chemical potential will increase in accord with Equation (A.3.1). The minus sign in the adsorption isotherm then serves to decrease the surface tension, as expected from experience.

Let us journey to understand the excess quantities in greater inwardness (thanks to Volker Heine for the phrase). The first term in

---

[4]More strictly, vacuum. We have to assert that the surface is in equilibrium with the underlying phase but not with the gas phase — otherwise we may not assume a single, α, phase.

the parentheses in (11.1.6) is the amount of entropy per unit area in the system, the layer, take away n_1/n_1^α times the total entropy contained in a chosen amount of homogeneous phase α containing n_1^α moles of our chosen component 1, see Equation (11.1.5). This leads us to the definition of the excess entropy as the difference between the amount of entropy of unit area of the layer and the amount of entropy of homogeneous phase having the same amount of component 1. Similarly, the excess, Γ_i, in (11.1.7a) is $1/A$ times the *number of moles of i in the layer*, take away the *number of moles of i in a piece of homogeneous phase, α, having the same number of moles of component 1 as are in the system, or layer*. Clearly, if $i = 1$, this is zero by construction so there is no excess of solvent. We can express the excess in general using Cahn's notation, and our rule that layer quantities are denoted using lower case sans serif font,

$$[Z/X] = \mathsf{z} - \frac{\mathsf{x}}{X^\alpha}Z^\alpha \qquad (11.1.9)$$

again: the amount of extensive quantity Z belonging to unit area of interphase compared to the amount of Z in homogeneous phase containing the same amount of extensive quantity X as the interphase. Hence this is the excess of Z with respect to X. It is helpful in view of what follows to point out that mathematically we may write

$$[Z/X] = \frac{1}{X^\alpha}\begin{vmatrix} \mathsf{z} & \mathsf{x} \\ Z^\alpha & X^\alpha \end{vmatrix} \qquad (11.1.10)$$

as you can see by expanding out the determinant. In view of this notation, I can write the excess in (11.1.7a) as

$$\Gamma_i = [n_i/n_1] \qquad (11.1.7b)$$

meaning the excess of species i with respect to species 1 (the solvent usually — if there is one, but we are not restricted by this). Alternatively speaking, it is the excess of species i having eliminated the change, $d\mu_1$, in the chemical potential of a chosen reference component. To be strict, we should indicate in the notation the constraint we have used: the right-hand side is unambiguous, but the left-hand side would be written more completely as, say, $\Gamma_i^{(1)}$. After all, I could have decided to eliminate the change in pressure, dp, from Equations (11.1.5) by multiplying the Gibbs–Duhem equation by v/V^α and subtracting from the Gibbs adsorption equation. The result, equivalent to (11.1.6), is

$$d\gamma = -\left(s - \frac{v}{V^\alpha}S^\alpha\right)dT - \left(n_i - \frac{v}{V^\alpha}n_i^\alpha\right)d\mu_i$$
$$= -[S/V]dT - [n_i/V]d\mu_i$$

This introduces an excess

$$\Gamma_i^{(V)} = [n_i/V] = \left(n_i - \frac{v}{V^\alpha}n_i^\alpha\right)$$

which is the difference between the amount of component i in unit area of interphase and the amount of component i in an equal volume of homogeneous phase. This is independent of the volume of the interphase and so is a perfectly legitimate measure of the amount of segregation which does not single out a particular component as the solvent (see Problem 11.2).

11.1.2   *The case of two phases*

As I have said, for the metal–electrolyte interphase, it is enough to consider the case of a single phase. Therefore, at a first reading, this section may be skipped without loss in the following chapters. However, for completeness, let us now pursue the instance when the components i are distributed in equilibrium in both phases, α and β, to left and right of the interface region in Figure 11.1(b). Take a deep breath, and read on.

By comparison with (11.1.5) and taking in view that there are now two phases, there are two Gibbs–Duhem equations and hence two constraints on the variations that may be made in the intensive quantities. We have these three equations:

$$d\gamma = -sdT + vdp - n_id\mu_i \qquad (11.1.11a)$$

$$0 = -S^\alpha dT + V^\alpha dp - n_i^\alpha d\mu_i \qquad (11.1.11b)$$

$$0 = -S^\beta dT + V^\beta dp - n_i^\beta d\mu_i \qquad (11.1.11c)$$

We label the extensive quantities in the homogeneous phases with superscripts α and β; we do not need to label T, p and the chemical potentials because they are the same for each component throughout the system in equilibrium. The two Gibbs–Duhem equations prescribe two constraints on the $C + 2$ intensive variables $\{T, p, \mu_i\}$; alternatively they may be solved to tell us how any two must evolve if we freely vary the others. Cahn analyses the three Equations (11.1.11) using the properties of determinants. First, following Finnis, write these as three simultaneous equations:

$$d\gamma = a_{11}x_1 + a_{12}x_2 + \cdots + a_{1\,C+2}\,x_{C+2}$$

$$0 = a_{21}x_1 + a_{22}x_2 + \cdots + a_{2\,C+2}\,x_{C+2}$$

$$0 = a_{31}x_1 + a_{32}x_2 + \cdots + a_{3\,C+2}\,x_{C+2} \qquad (11.1.12)$$

where x stands for the differential of one of the $C + 2$ intensive variables and a stands for one of the $3C + 6$ extensive quantities $\{s, v, n_i, S^\alpha, V^\alpha, n_i^\alpha, S^\beta, V^\beta, n_i^\beta\}$. We decide to eliminate derivatives of two of the intensive variables, $\{dT, dp, d\mu_i\}$, represented by x_ℓ and x_k. First form a matrix

$$c = \begin{pmatrix} c_{11} \; a_{1\ell} \; a_{1k} \\ c_{21} \; a_{2\ell} \; a_{2k} \\ c_{31} \; a_{3\ell} \; a_{3k} \end{pmatrix}$$

whose entries c are formed from one of the other columns of Equations (11.1.12). The determinant of c is constructed from its cofactors:

$$\det c = \begin{vmatrix} c_{11} \; a_{1\ell} \; a_{1k} \\ c_{21} \; a_{2\ell} \; a_{2k} \\ c_{31} \; a_{3\ell} \; a_{3k} \end{vmatrix} = c_{11}C_{11} + c_{21}C_{21} + c_{31}C_{31}$$

where the cofactors are

$$C_{11} = a_{2\ell}a_{3k} - a_{2k}a_{3\ell}; \quad C_{21} = -\left(a_{1\ell}a_{3k} - a_{1k}a_{3\ell}\right);$$

$$C_{31} = a_{1\ell}a_{2k} - a_{1k}a_{2\ell}$$

I now multiply each of Equations (11.1.12) in turn by one of the cofactors:

$$C_{11}d\gamma = \left(a_{2\ell}a_{3k} - a_{2k}a_{3\ell}\right)\left(a_{11}x_1 + a_{12}x_2 + \cdots + a_{1\,C+2}x_{C+2}\right)$$

$$0 = -\left(a_{1\ell}a_{3k} - a_{1k}a_{3\ell}\right)\left(a_{21}x_1 + a_{22}x_2 + \cdots + a_{2\,C+2}x_{C+2}\right)$$

$$0 = \left(a_{1\ell}a_{2k} - a_{1k}a_{2\ell}\right)\left(a_{31}x_1 + a_{32}x_2 + \cdots + a_{3\,C+2}x_{C+2}\right)$$

Add these three together and expand them out, and consider just the three terms in x_1:

$$C_{11}d\gamma = (a_{2\ell}a_{3k} - a_{2k}a_{3\ell})\,a_{11}x_1 - (a_{1\ell}a_{3k} - a_{1k}a_{3\ell})\,a_{21}x_1$$
$$+ (a_{1\ell}a_{2k} - a_{1k}a_{2\ell})\,a_{31}x_1 + \cdots$$

$$= \begin{vmatrix} a_{11} & a_{1\ell} & a_{1k} \\ a_{21} & a_{2\ell} & a_{2k} \\ a_{31} & a_{3\ell} & a_{3k} \end{vmatrix} x_1 + \cdots$$

and the second equality follows from the same cofactor property of determinants as above. The same will hold if I single out the terms in x_2, x_3 and so on, so it follows that

$$C_{11}d\gamma = \sum_{n=1}^{C+2} D_n x_n \qquad (11.1.13)$$

where D_n is the determinant:

$$D_n = \begin{vmatrix} a_{1n} & a_{1\ell} & a_{1k} \\ a_{2n} & a_{2\ell} & a_{2k} \\ a_{3n} & a_{3\ell} & a_{3k} \end{vmatrix} = a_{1n}C_{11} + a_{2n}C_{21} + a_{3n}C_{31}$$

This is the equivalent of (11.1.6) after we multiplied (11.1.5) by a factor and subtracted it from the adsorption equation. Now, when there are two phases, we have taken an appropriate linear combination of (11.1.11b) and (11.1.11c) and added it to (11.1.11a). In the case of one phase, we eliminated one of the chemical potentials; now we have two intensive variables to eliminate; let's choose $x_\ell = d\mu_1$ and $x_k = d\mu_2$.[5] With a little thought, by comparison with (11.1.12), you can see that in this case,

$$a_{1\ell} = n_1$$
$$a_{2\ell} = n_1^\alpha$$
$$a_{3\ell} = n_1^\beta$$

[5]You don't *have* to choose 1 and 2. If you want, you could use, say, m and n for 1 and 2 in what follows, but it's up to me how I number my components so I may as well place the ones I want to eliminate at the top of the list!

$$a_{1k} = n_2$$

$$a_{2k} = n_2^\alpha$$

$$a_{3k} = n_2^\beta$$

and substituting the actual names of variables into (11.1.13), we have

$$C_{11} = n_1^\alpha n_2^\beta - n_2^\alpha n_1^\beta = \begin{vmatrix} n_1^\alpha & n_2^\alpha \\ n_1^\beta & n_2^\beta \end{vmatrix}$$

and

$$d\gamma = -\frac{\begin{vmatrix} s & n_1 & n_2 \\ S^\alpha & n_1^\alpha & n_2^\alpha \\ S^\beta & n_1^\beta & n_2^\beta \end{vmatrix}}{\begin{vmatrix} n_1^\alpha & n_2^\alpha \\ n_1^\beta & n_2^\beta \end{vmatrix}}dT + \frac{\begin{vmatrix} v & n_1 & n_2 \\ V^\alpha & n_1^\alpha & n_2^\alpha \\ V^\beta & n_1^\beta & n_2^\beta \end{vmatrix}}{\begin{vmatrix} n_1^\alpha & n_2^\alpha \\ n_1^\beta & n_2^\beta \end{vmatrix}}dp - \frac{\begin{vmatrix} n_i & n_1 & n_2 \\ n_i^\alpha & n_1^\alpha & n_2^\alpha \\ n_i^\beta & n_1^\beta & n_2^\beta \end{vmatrix}}{\begin{vmatrix} n_1^\alpha & n_2^\alpha \\ n_1^\beta & n_2^\beta \end{vmatrix}}d\mu_i \quad (11.1.14)$$

This equation is rendered more compactly by Cahn's notation. But before showing that, consider the following three points.

(1) The multiplier to each $d\mu_i$ is an excess concentration with respect to two components 1 and 2 (cf. (11.1.9) and (11.1.10) expressed as 2×2 determinants divided by "1×1 determinants").
(2) The implicit sum over all components remains in the third term as it does in (11.1.6), but while in that case the first term, $i = 1$, vanished; now the *first two* terms, $i = 1$ and $i = 2$, vanish by a property of a determinant, namely that it is zero if any two rows or columns are identical.
(3) A second property of determinants guarantees that the numerators in (11.1.14) are independent of the width of the layer. The value of a determinant is unchanged if a multiple of one row (or column) is added to another row (or column). If I move the left-hand side vertical line in Figure 11.1(b) to the left, then since this lies in homogeneous phase, I am adding a multiple of the second row to the first row or of the third row to the first if I move the right-hand side vertical line to the right.

In fact all the ratios of determinants in (11.1.14) are *excess* extensive quantities. Cahn extends the notation of (11.1.10) to allow the elimination of two variables:

$$[Z/XY] = \frac{\begin{vmatrix} z & x & y \\ Z^\alpha & X^\alpha & Y^\alpha \\ Z^\beta & X^\beta & Y^\beta \end{vmatrix}}{\begin{vmatrix} X^\alpha & Y^\alpha \\ X^\beta & Y^\beta \end{vmatrix}} \qquad (11.1.15)$$

In Cahn's words, "The physical meaning of $[Z/XY]$ is the difference in the amount of Z between a unit area of the layer and two portions of the homogeneous phase having the same total amount of X and Y as the layer." He means, one portion each of the homogeneous phases α and β. Now we can write (11.1.14) more neatly:

$$d\gamma = -[S/n_1 n_2]\, dT + [V/n_1 n_2]\, dp - [n_i/n_1 n_2]\, d\mu_i \qquad (11.1.16)$$

The sum over i is still implicit, but by construction, $[n_1/n_1 n_2]$ and $[n_2/n_1 n_2]$ are zero as observed earlier. For this case, our excess concentration is

$$\Gamma_i = [n_i/n_1 n_2] \qquad (11.1.7c)$$

and the Gibbs adsorption isotherm becomes

$$\left(\frac{\partial\gamma}{\partial\mu_i}\right)_{T,p,\mu_j(j>2,j\neq i)} = -\Gamma_i \quad (i>2) \qquad (11.1.17)$$

This fixes both of the errors associated with Equation (11.1.3b). First, in the left-hand side, the variation can be made without violating the phase rule because two of the intensive variables, μ_1 and μ_2, will adjust themselves in order to maintain equilibrium and we do not insist that they are to remain constant. Second, the right-hand side is independent of the width of the layer as proved in point 3. The interfacial excess in (11.1.17) is not the same as the Γ_i in Equation (11.1.7b) — as I have pointed out, Cahn's notation is unambiguous, and really we should decorate our Γs with suitable superscripts. In fact, the symbol Γ was used by Gibbs himself for the excess; his Γ corresponds to our n. Cahn points out that his extension has a number of benefits compared to Gibbs's treatment; one of which is that the "volume", v, of Gibbs's interface is zero and indeed Gibbs regards the interface as a mathematical "dividing surface". Another benefit is that for Gibbs, but not necessarily for Cahn, one of the X or Y in the excess (11.1.15) *must* be volume. Let's pursue that a bit. Suppose I would like to eliminate $d\mu_1$ and dp. Then my excess is

$$\Gamma_i = [n_i/n_1 V] = \frac{\begin{vmatrix} n_i & n_1 & v \\ n_i^\alpha & n_1^\alpha & V^\alpha \\ n_i^\beta & n_1^\beta & V^\beta \end{vmatrix}}{\begin{vmatrix} n_1^\alpha & V^\alpha \\ n_1^\beta & V^\beta \end{vmatrix}} \qquad (11.1.7d)$$

Gibbs called this excess $\Gamma_{i(1)}$. If you set $n_i^\beta = 0$, $\forall i$, and then work out the determinants, you'll find

$$\Gamma_i = n_i - \frac{n_1}{n_1} n_i^\alpha$$

which is the same as (11.1.7a). This is to be expected because $n_i^\beta = 0$ indicates that there are no components in the β-phase. This would be the case for the electrolyte–electrode interface if the components 1 and 2 are not to be found in the metal.

I should point out that Equation (11.1.16) becomes absurd if there are only two components since the dependence of the surface tension on the composition vanishes. So if there are two phases and two components, then an admissible excess of component 2 is $[n_2/n_1 V]$ or of component 1 is $[n_1/n_2 V]$. Finnis has shown that this excess is identical to the excess defined in terms of Gibbs's dividing surface.

11.2 The Electrocapillary and Lippmann Equations

The Lippmann equation states how the interfacial tension of an electrode depends on its Galvani potential. The interest arises from experiments made nearly 150 years ago on mercury electrodes. Almost all such experiments are made using mercury because this is the only pure liquid metal at room temperature (and amalgams can be used to study effects of alloying) and hence the only metal whose surface tension can be measured using its capillarity. In an electrochemistry experiment, the surface tension of mercury can be measured as a function of the composition of the electrolyte with which it is in contact and its Galvani potential compared to a standard potential. Important results which are derived in this chapter are the electrocapillary equation:

$$d\gamma = -\Gamma_i d\mu_i - \sigma \, d\phi$$

where σ is the surface charge density on the metal electrode, which relates the change in interfacial tension to changes in composition and inner potential, and from which follows the Lippmann equation:

$$\left(\frac{\partial \gamma}{\partial \phi}\right)_\mu = -\sigma$$

and the result

$$\Gamma_i = \left(\frac{\partial \gamma}{\partial \mu_i}\right)_{\phi, \mu_{j \neq i}}$$

which permits measurement of the relative adsorption of species to the metal surface. An important measurable quantity is the *potential of zero charge* (PZC) which is the value of ϕ for which the left-hand side of the Lippmann equation is zero. Details of how the measurements are made can be found in Schmickler and Santos, and Bockris and Reddy. Unless the electrode is liquid Hg, it's not possible to measure the surface tension directly. However, a relative surface tension can be measured by integration of the surface charge over the electric potential, starting at a reference potential more negative than the PZC:

$$\gamma(\phi) - \gamma(\phi_{\text{ref}}) = \int_{\phi_{\text{ref}}}^{\phi} \sigma(\phi')d\phi'$$

while the surface charge density, σ, is measured by integrating from the PZC (chronocoulometric (!) method) or integration over the capacity — see Equation (11.2.5).

There are at least two ways to derive the Lippmann equation and it will be worthwhile to include two here. A further key point is that the Lippmann equation permits the mapping of the electric properties of the interphase to a flat plate capacitor of capacity, $\mathcal{C}$ [μF cm^{-2}], which may or may not be a function of the Galvani potential. The measured dependence of $\mathcal{C}$ upon ϕ provides information on the extent of adsorption of ions from the electrolyte on the metal electrode surface.

11.2.1 *Derivation number one*

Recall the electrochemical cell in Figure 2.1. Metal which is dipped into electrolyte may become charged. Imagine that Figure 11.1(b) represents a section through the metal–electrolyte interphase. The system enclosed within the boundary in Figure 11.1(b) is electrically neutral. I will re-derive Equation (11.1.3a). The combined first and second laws read:

$$dU = TdS - pdV + \mu_i dn_i + \gamma \, dA + \phi \, dq_m \qquad (11.2.1)$$

which is the same as (11.1.1) but there is one additional work term. If I increase the electrode charge by dq_m, then I carry that charge from the electrolyte at potential ϕ_s to the metal at potential ϕ_m so I do an amount of work $\phi \, dq_m$ on the system if $\phi = \phi_m - \phi_s$. As usual, we integrate (11.2.1):

$$U = TS - pV + \mu_i n_i + \gamma A + \phi q_m$$

Therefore,

$$\gamma A = U - TS + pV - \mu_i n_i - \phi q_m$$

and taking the total differential and inserting (11.2.1) in place of dU,

$$d\left(\gamma A\right) = \gamma \, dA + A d\gamma = \gamma \, dA - SdT + V dp - n_i d\mu_i - q_m \, d\phi$$

so that we have $d\gamma$ in terms of variations in intensive quantities only:

$$A d\gamma = -SdT + V dp - n_i d\mu_i - q_m \, d\phi$$

We are allowed to construct excesses assuming a single phase as in (11.1.6) because none of the species i reside in the metal electrode. Therefore, the Gibbs adsorption equation in the case of a charged electrode is

$$d\gamma = -\left[S/n_1\right] dT + \left[V/n_1\right] dp - \left[n_i/n_1\right] d\mu_i - \sigma \, d\phi$$

and at constant T and p, we have

$$d\gamma = -\Gamma_i d\mu_i - \sigma \, d\phi \qquad (11.2.2)$$

where Γ_i is given by (11.1.7b) and $\sigma = q_m/A$ is the electrode surface charge density. This is called the *electrocapillary equation*. There is no difficulty in asserting that σ is an unambiguous excess interfacial charge because the amount of charge transferred is exactly equal to minus the excess of electrons compared to the neutral metal and these all reside in a uniform sheet at the electrode surface. I repeat that ϕ is the Galvani or inner potential difference across the electrode–electrolyte interface and this is of course not measurable. However, as in Figure 2.4, we can imagine that the cell is completed by connecting to a perfectly reversible, non-polarisable standard hydrogen electrode. If the voltmeter is replaced by a source of electricity delivering an applied e.m.f., $\mathcal{E}_a$ volts, and if its negative terminal is connected to the SHE,

$$\mathcal{E}_a = (\phi_{Cu} - \phi_{Cu'})$$
$$= (\phi_{Cu} - \phi_m) + (\phi_m - \phi_s) + (\phi_s - \phi_{Pt}) + (\phi_{Pt} - \phi_{Cu'})$$

To be more general, I have replaced Zn with an unspecified electrode metal with inner electric potential ϕ_m. The metal–metal inner potential differences are all differences in Fermi energy (see Equation (3.2.1)) and therefore constant. Then as I vary the applied e.m.f.,

$$d\mathcal{E}_a = d(\phi_s - \phi_{Pt}) + d(\phi_m - \phi_s)$$
$$= d(\phi_s - \phi_{Pt}) + d\phi$$

I multiply through by σ and insert into (11.2.2),

$$d\gamma = -\sigma\, d\mathcal{E}_a - \Gamma_i d\mu_i - \sigma\, d(\phi_{Pt} - \phi_s)$$

For the last term on the right-hand side, we can use the Nernst equation (9.2.1):

$$d(\phi_{Pt} - \phi_s) = \frac{RT}{F} d\ln a_{H^+} = \frac{RT}{F} d\mu_{H^+}$$

where a_{H^+} and μ_{H^+} are the activity and chemical potential of the proton that carries the current leakage across the non-polarisable electrode interphase. F is the Faraday constant. Finally, then,

$$d\gamma = -\sigma\, d\mathcal{E}_a - \Gamma_i d\mu_i - \sigma\frac{RT}{F} d\mu_{H^+} \qquad (11.2.3)$$

In an experiment, the concentration of electrolyte and activity of hydrogen ions in the SHE will be controlled and held constant and so (11.2.3) becomes a form of the Lippmann equation:

$$\left(\frac{\partial \gamma}{\partial \mathcal{E}_a}\right)_\mu = -\sigma \qquad (11.2.4)$$

If I differentiate a second time,

$$\left(\frac{\partial^2 \gamma}{\partial \mathcal{E}_a^2}\right)_\mu = -\left(\frac{\partial \sigma}{\partial \mathcal{E}_a}\right)_\mu = -\mathcal{C} \qquad (11.2.5)$$

this defines the "differential capacity" of the interphase between electrode and electrolyte. Unless we study the molecular structure in detail, we can only assert that the layer of electrons at the metal surface and the layer of equal and opposite charge somewhere in the solution are modelled as a flat plate capacitor with capacity $\mathcal{C}$. Because this must be positive, a graph of γ versus $\mathcal{E}_a$ must have the shape of a parabola (if $\mathcal{C}$ is constant) having a *maximum* at $\sigma = 0$, see Figure 11.2. The potential, ϕ_{pzc}, at the maximum is *the potential of zero charge*, PZC.

Table 11.1 shows some potentials of zero charge, in V/SHE, for metals in a 0.01 molal solution of either potassium chloride or sodium sulphate.

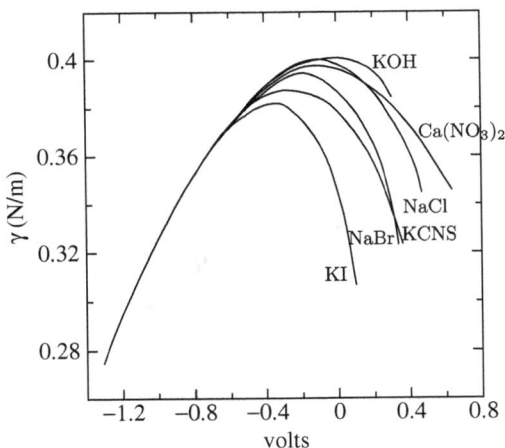

Fig. 11.2. Interfacial tension of mercury in contact with aqueous solutions of a number of salts (KCNS is potassium thiocyanate). The potential scale in the abscissa is somewhat arbitrary.
Source: Adapted with permission from David C. Grahame in *Chem. Rev.*, **41**, 441 (1947).

Table 11.1.   Potential of zero charge in either potassium chloride or sodium sulphate solutions.

Al	Bi	Cd	Pb	Co	Cu	Au	Ag
KCl	KCl	KCl	KCl	Na_2SO_4	Na_2SO_4	Na_2SO_4	Na_2SO_4
-0.52	-0.36	-0.92	-0.69	-0.32	$+0.03$	$+0.23$	-0.70

11.2.2   *Derivation number two*

Since we already have the Lippmann equation, this section can be skipped on a first reading. But it is instructive to follow this second derivation because it does not require the additional work term in (11.2.1). Instead the charge and potential are allowed to appear as a consequence of using *electro*chemical potential in the Gibbs adsorption isotherm. The picture that we need of the interphase requires us to imagine the metal phase to be inert and impenetrable to the molecules and ions in the electrolyte, so these all have excesses, Γ_i, as given in (11.1.7b). But the metal is dissolving into its ions and so the equilibrium

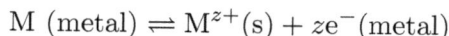

$$\text{M (metal)} \rightleftharpoons \text{M}^{z+}(\text{s}) + z\text{e}^-(\text{metal})$$

is established in the interphase. In terms of electrochemical potential, the Gibbs adsorption isotherm is

$$d\gamma = -\Gamma_i d\tilde{\mu}_i - \Gamma_e d\tilde{\mu}_e \qquad (11.2.6)$$

The index i labels all of the species in solution, including M^{z+}, charged or not; $i = 1$ refers to water, and excesses, Γ_i, are relative to solvent. The "excess" of electrons per unit area is not controversial since this is clearly the number of moles of electrons over and above those that make the metal neutral and are all confined to a sheet at the surface of the electrode. I have omitted the contribution from the "excess" of metal, $-\Gamma_M d\mu_M$, because the chemical potential of a pure metal is constant (standard state) $d\mu_M = 0$. We now break up (11.2.6) into its terms. The second term is, in view of (3.4.4),

$$-\Gamma_e d\tilde{\mu}_e = -\Gamma_e d\mu_e + F\Gamma_e \, d\phi_m = -\sigma \, d\phi_m$$

because μ_e is the Fermi level and is constant, independent of the inner potential (see Figure 3.1), so $d\mu_e = 0$. $-F\Gamma_e = \sigma$ is the excess charge per unit area on the surface of the metal in contact with the solution. The first term on the right-hand side of (11.2.6) is, in accord with (4.1.1),

$$-\Gamma_i d\tilde{\mu}_i = -\Gamma_i d\mu_i - F\Gamma_i z_i \, d\phi_s$$
$$= -\Gamma_i d\mu_i + \sigma \, d\phi_s$$

where z_i are the charge numbers of species i. The second line follows because the total excess charge in the electrolyte is just the sum of excesses of all the charged ionic species including M^{z+}, and this must equal σ to guarantee overall charge neutrality. In fact, these are the ions that are accumulated in the double layer, attracted to the charged metal surface. Hence,

$$\sigma = -F\Gamma_i z_i$$

and we can retain the sum over all i because uncharged species have $z_i = 0$ and won't contribute to the sum. (11.2.6) now becomes

$$d\gamma = -\sigma \, d\phi - \Gamma_i d\mu_i$$

where $\phi = \phi_m - \phi_s$ is the Galvani potential difference between electrolyte and electrode. Once again, this leads us to the Lippmann equation:

$$\left(\frac{\partial \gamma}{\partial \phi}\right)_\mu = -\sigma \qquad (11.2.7)$$

this time without making contact with a reference electrode. Here, and in (11.2.4) and (11.2.5), we must be careful with the "constant μ" subscript. I have been deliberately vague. Not all the chemical potentials can be held constant as we saw in Section 11.1. So I mean the partial derivative to be taken with all chemical potentials being constant that can be held fixed. In the case that we regard the metal phase as being inert, this means the excesses are given by (11.1.7a) and the derivative of the solvent chemical potential is eliminated and cannot be controlled.

11.3   Measurement

The principal difference in comparing derivations number one and two are in how the charges and potential turn up in the combined first and second law — first as an explicit work term and second by writing the adsorption isotherm in terms of electrochemical rather than chemical potential. The first derivation, which I think was Lippmann's original in outline, is based purely in macroscopic thermodynamics, whereas the second has a closer focus on the ions in the electrolyte; in fact, we included explicitly the presence of those ions which appear as a result of corrosion of the electrode. These could, of course, be absent, as they are in most electrocapillary experiments using mercury.

An example of such an experiment is shown in Figure 11.2. It is typical that when the electrode potential is most negative and the electrode is negatively charged, the curve is close to parabolic and is the same for all salts. This is because in this condition, it is the *cation* that is attracted to the surface of the metal. Conversely when the metal is positively charged, one speaks of "specific adsorption" and the structure of the anions in the double layer causes a variation in the capacitance. The explanation usually given is that the solvation of a cation is much stronger than that of an anion. This is because the cation is small and so is easily surrounded by water molecules. Therefore, the cation is protected by its solvation shell from being physically adsorbed on the metal surface. The anion, on the other hand, is more easily freed from its solvation shell and so adsorbs rather strongly.[6] This is clearly observed in impedance spectroscopy and cyclic voltammetry. An example is shown in Figure 11.3. The abscissa shows the voltage relative to the standard calomel electrode (Section A.3.6). The two upper figures show the capacity in the ordinate. The two lower figures are *cyclic voltammograms*, showing the current density as a function of voltage. The electrode is a single crystal silver (111) surface; the left-hand side figures are measurements

---

[6]Schmickler and Santos make much mention of "inner" and "outer" sphere reactions. I think that inner sphere refers to the case in which the ion is not completely screened by its solvation shell, i.e., it is adsorbed at the electrode, and outer sphere reactions involve a fully solvated ion, presumably some distance from the electrode, in the electrolyte.

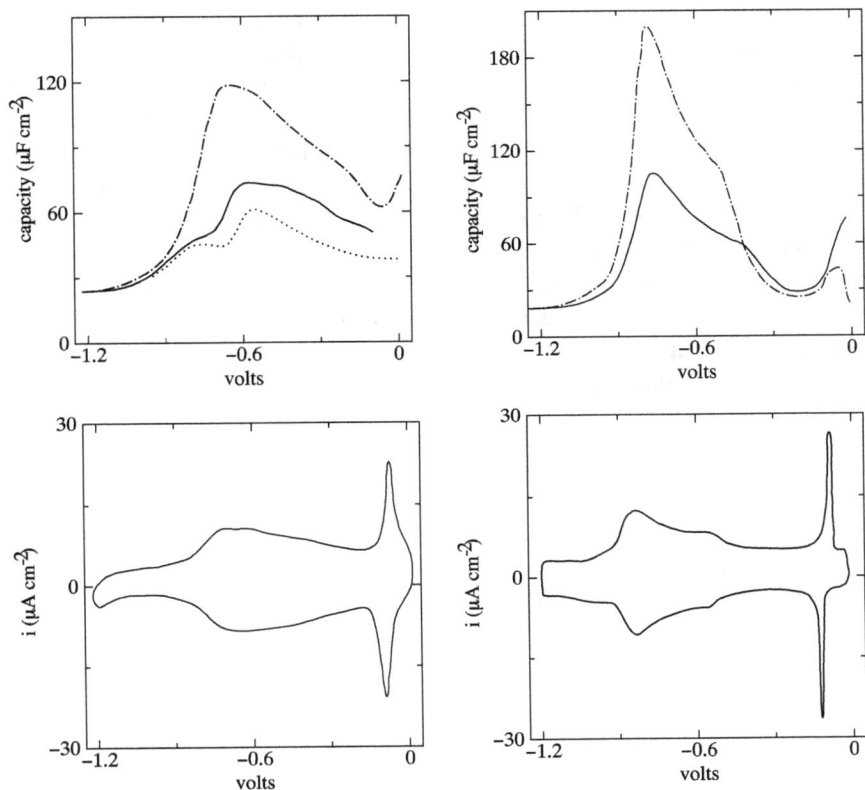

Fig. 11.3. The upper two figures show differential capacity at 20 Hz versus potential (standard calomel scale) for Cl$^-$ (left) and Br$^-$ (right) ions adsorbed on the (111) surfaces of single crystals of silver. Broken and solid curves respectively correspond to ion concentrations of 10 and 0.1 millimolar; the dotted line is at zero concentration. The two lower curves are cyclic voltammograms (essentially i–V curves) at 10 nM concentrations.

Source: Adapted with permission from G. Beltramo and Elizabeth Santos, *J. Electroanalytical Chem.*, **556**, 127 (2003).

in chloride electrolyte and the right-hand side figures in bromide. The solid and broken lines correspond to concentrations of 0.1 millimolar and 10 nM, respectively. The dotted line is in the absence of chloride or bromide. The upper curves are examples of "impedance spectroscopy": the voltage is increased at a rate of 10 mV s^{-1} and is impressed over an alternating voltage of 20 Hz. To me, the experiment is reminiscent of the LCR circuit which is the electrical analogue of a typical resonant system and comprises a resistor, a capacitor and a choke in series. The differential capacity–potential curves in this experiment are modelled by a simple equivalent circuit of a resistor and capacitor *in series* (not in parallel as in Figure 2.6) which represents the electrolyte by a resistor and the polarisable interphase by a capacitor. The paper by Beltramo and Santos is well worth reading. The capacity–potential curve is discussed further in the following section. For brevity, I will just state here that the interpretation of the broad and narrow peaks in the cyclic voltammograms is interpreted as a result of the growth of disordered and ordered adsorbed anions on the metal surface, and the ordered structure is confirmed by scanning tunnelling microscopy experiments.

Chapter 12

Atomistic Models of the Interphase and Interphase Capacitance

It is the experimental observation that the electric properties of the electrode–electrolyte interphase may be mapped to an equivalent circuit that suggests to us that the *ideally polarisable* interphase may be modelled as a flat plate capacitor. If we then ponder on what the atomic arrangement of ions and solvent may look like, we imagine, in the first instance, a *Helmholtz double layer* comprising an ordered layer of ions attached to the metal surface, just of sufficient concentration so as to neutralise the layer of electrons or holes residing at the surface of the electrode. This in fact, as mentioned earlier, can be observed by STM and possibly low energy electron diffraction and deduced by the peaks in the cyclic voltammogram in Figure 11.3.

Figure 12.1 is adapted from Bockris and Reddy and shows an insightful cartoon into how the double layer and the interphase may look. Note how the cations are fully solvated (outer sphere) while the anions are adsorbed (inner sphere). However, it is not sufficient to expect a static double (or triple) layer. An accumulation of ions in solvent is bound to be to some extent disordered at finite temperature. The model for this interphase was put forward by Gouy and by Chapman as early as 1910.

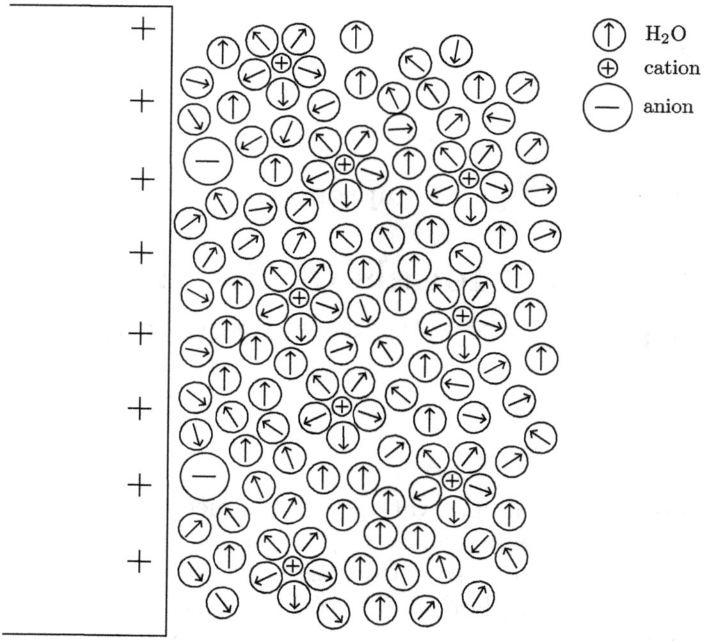

Fig. 12.1. A cartoon to represent the atomic scale structure of the interphase. To the left is the electrode, in this case, positively charged with its charge all concentrated into a surface sheet (since the metal cannot support an electric field in the absence of current). This sheet of charge can be reasonably modelled as the plate of a capacitor. In the electrolyte are three species: water, anion and cation. Water has the following two remarkable properties. (*i*) It has a large dipole moment (indicated by an arrow) having a positive end (two protons) and a negative end (the oxygen ion). This means that it can electrically *screen* both positive and negative charges. It is seen to be doing this to screen the positive charge of the cation, whose electric field then can't been seen by a test charge a distance away. (*ii*) It can form *hydrogen bonds* so as to complete a four-fold coordinated network of molecules (this is missing in the cartoon). Cations are usually quite small metal ions and can be easily screened by water molecules. Anions tend to be large, for example, SO_4^{--}, Cl^-, and it is harder for them to be screened; in the cartoon, some anions are absorbed on the surface of the electrode. It is less obvious that the charge within the electrolyte near the electrode can be modelled as a flat sheet, but if we accept that, then we have the picture of the double layer as a flat plate capacitor. It is worthwhile to think a bit about how such a structure will respond to changes in electric field, in particular an oscillating field as in the experiment whose results are shown in Figure 11.3.

Source: Adapted with permission from Bockris and Reddy (see Further Reading in Appendix D).

12.1   Gouy–Chapman Theory

The model is of point ions solvated by a continuum dielectric of permittivity ϵ, distributed according to the principles of electrostatics and statistical mechanics. In the simplest case of a planar electrode and a one-to-one electrolyte (such as NaCl), we place the electrode surface at the coordinate $x = 0$ and consider a one-dimensional distribution of the electric potential, $\phi(x)$, which is given in terms of the charge density $\rho(x)$ by the Poisson equation:

$$\frac{d^2\phi}{dx^2} = -\frac{\rho(x)}{\epsilon}$$

The point ions may be either cations or anions and their densities will be denoted $n_+(x)$ and $n_-(x)$, respectively. Hence the total charge density is

$$\rho(x) = ze\left(n_+(x) - n_-(x)\right)$$

where z is the charge number and e is the proton elementary charge. As mentioned earlier, we assume $z = 1$ for simplicity but z can be easily retained in the more general case. If we set the zero of electric potential at $x = \infty$, then according to the Boltzmann statistics,

$$n_+ = n_0 e^{-e\phi(x)/kT}$$

$$n_- = n_0 e^{+e\phi(x)/kT}$$

where n_0 is the bulk concentration of cations and anions, $n_0 = n_+(\infty) = n_-(\infty)$. The resulting Poisson–Boltzmann equation is

$$\frac{d^2\phi}{dx^2} = -\frac{en_0}{\epsilon}\left(e^{-e\phi(x)/kT} - e^{+e\phi(x)/kT}\right)$$

A transparent solution exists if the Poisson–Boltzmann equation is linearised, that is, in the limit that $e\phi(x)/kT \ll 1$, namely

$$\frac{d^2\phi}{dx^2} = \kappa^2\phi(x)$$

where

$$\kappa = \lambda^{-1} = \sqrt{\frac{2e^2n_0}{\epsilon kT}}$$

Table 12.1. Debye length, in ångstrom as a function of molar concentration.

Molar conc^n	10^{-4}	10^{-3}	10^{-2}	0.1
$\lambda/\text{Å}$	304	96	30.4	9.6

and λ is the *Debye length*. Table 12.1 shows Debye lengths for a one-to-one electrolyte at room temperature.

The solution to the linearised Poisson–Boltzmann equation, subject to the boundary condition, $\phi(x) \to 0$ as $x \to \infty$ is

$$\phi(x) = A e^{-\kappa x}$$

and the constant A may be established by equating the total charge density from the anions and cations existing between $x = 0$ and $x = \infty$ to minus the charge density at the surface of the electrode in order to ensure overall charge neutrality:

$$\int_0^\infty \rho(x)\mathrm{d}x = -\sigma$$

Combining this with the Poisson equation results in the following two formulas:

$$\phi(x) = \frac{\sigma}{\epsilon\kappa} e^{-\kappa x}$$

$$\rho(x) = -\sigma\kappa\, e^{-\kappa x}$$

$\rho(x)$ describes a layer of *space charge* which exactly balances the charge on the electrode.

This leads us back to the notion that the interphase acts as a parallel plate capacitor. The capacitance is charge over voltage at the interface, $x = 0$, or equivalently ϵ times area divided by the distance between the plates. In this case, we have a *capacity* defined as capacitance per unit area and this is clearly

$$C = \frac{\sigma}{\phi(0)} = \epsilon\kappa \quad [\mu\text{F cm}^{-2}]$$

a nice simple formula which identifies the "plate separation" of the equivalent circuit capacitor (Section 11.3 and Figure 12.1) as the

Debye length. Since this is typically a few ångstroms, the capacity is quite large.

While the linearised Poisson–Boltzmann equation gives useful insight into the equivalent capacitor, it is really necessary to use the solution from the non-linearised equation. The maths is not very edifying and is given by Schmickler and Santos and by Bockris and Reddy (see Further Reading in Appendix D). The result for the double layer capacity is

$$C = \epsilon\kappa \, \cosh \frac{e\phi(0)}{2kT}$$

However, the electric potential at the origin cannot be measured. On the other hand, this potential differs by a constant from the inner, or Galvani, potential of the electrode. When $\phi(0) = 0$, then there is no charge on the electrode and hence its potential is the potential of zero charge (PZC), ϕ_{pzc}. So the Gouy–Chapman capacity is

$$C = \epsilon\kappa \, \cosh \frac{e(\phi - \phi_{\text{pzc}})}{2kT}$$

The cosh function has the shape of a *catenary* — a chain hanging under its own weight. Hence the capacity in the Gouy–Chapman theory is a simple convex function with a minimum at the PZC. This is not really observed in experiments. The upper curves in Figure 11.3 do show minima at voltages which are probably close to the PZC, but there are additional maxima and minima which are not described by the Gouy–Chapman capacity. To overcome this deviation from Gouy–Chapman capacity at high ion concentrations, the interphase is modelled as two capacitors in series, having respectively the Gouy–Chapman capacity, C_{GC}, and the *Helmholtz capacity*, C_{H}. Hence the interphase has capacity, C, given by

$$\frac{1}{C} = \frac{1}{C_{\text{GC}}} + \frac{1}{C_{\text{H}}}$$

Chapter 13

Kinetics

13.1 Corrosion

Up to now we have confined ourselves to the considerations of electrochemistry *in equilibrium*. Probably the key finding is the *electrochemical series*, Section 9.4, and the observation that *all metals* with
the exception of gold are thermodynamically unstable with respect
to their oxides (see Section 9.1). From this there arise two questions
as posed neatly by J. M. West (see Further Reading in Appendix D).
The first is, given this fact, why do not all metals spontaneously
waste away? I already answered this question in Chapter 1. If a piece
of metal is immersed in water, then some metal atoms will be converted to cations and migrate into the electrolyte, but *they will leave
behind their electrons* and cause the metal to become charged; this
accumulated charge will resist further dissolution. John West's next
question is more challenging: in that case, why do metals corrode at
all? There are several answers to this. One is that in the case that
a piece of metal is immersed in electrolyte and come to equilibrium,
then this state is achieved at a certain single electrode potential,
$^m\Delta^s\phi = \Delta\phi_{eq}$; if the electric potential within the metal can be con-
trived to be raised, then the equilibrium condition is shifted and
more metal can dissolve. Alternatively, the metal may be connected
electrically to an electron drain such as in an electrochemical cell.
In real-life corrosion, what mostly happens is that the macroscopic
surface of the metal divides itself into patches: some anodes and
some cathodes at which oxidation and reduction reactions take place
simultaneously, the electrons and ions flowing within the metal and

the electrolyte such as to complete the circuit and allow the anodic patches to waste away. This leads to the notion of a *mixed potential* which we come to in Chapter 16. In some particularly unfortunate cases, a base metal may be electrically connected to a more noble metal in order to create a natural electrochemical cell such as in the lemon lamp (Chapter 1). This really does happen, usually as consequence of poor design, as in the union of an aluminium car body with a steel chassis leg, for instance, or the use of fasteners and components of dissimilar metals (try to remove the bumper from a Jaguar XJ8 and you will find a demonstration of this probably accompanied by oaths and bruised knuckles as the fasteners will be seized to the brackets by corrosion).

Now follows the most crucial question of all from a practical point of view. How fast does a piece of metal corrode? In this text, I can only cover *uniform corrosion*. We are then interested in the uniform rate of metal loss, measured in, say, mm per year. By coincidence, the rate in mm per year is almost numerically the same as the corrosion current density in amps per metre squared. Again in real life, uniform corrosion is not of paramount interest since it can be designed against simply by specifying a life time based on, say, the wall thickness of a pipe. What is dangerous is *localised attack*. In the example in which the surface of the metal has negotiated its division into anodes and cathodes, the conservation of charge is expressed as the equality between the cathode and anode currents. If the anode areas are much less than the cathode areas, then the anode current density is much greater than the cathode current density, which leads to *pitting* holes appearing in the surface in which the electrolyte composition is vastly different compared to the bulk. When this is exacerbated by the presence of a chloride ion, the results will be catastrophic. Tony Entwisle used to add to the metallurgical dictum, "anodes are positive and they corrode", the statement, "if you investigate a corrosion failure, look for where the chloride came from".

However, I do not wish to wander further into the engineering aspects; let's continue to look at the physics of uniform corrosion — specifically, can we formulate its *rate*? This means pedagogically that we move from the subject of chemical thermodynamics to chemical kinetics. Our first aim is to take the steps needed to derive the famous Butler–Volmer equation. In the spirit of this text, I will assume a knowledge of some physics, in particular some statistical mechanics and solid state physics,[1] but no chemistry to speak of. The over-arching framework is the "transition state theory" that is foremost associated with the name of Henry Eyring who flourished in the 1930s. In the transition state theory, we talk of a rate coefficient, k_1 (usually called a "rate constant", but it's hardly constant, having an exponential dependence on temperature), in units of s^{-1}, which is described by the formula

$$k_1 = \frac{kT}{h}e^{-\Delta G^{\ddagger}/RT} \qquad (13.1.1)$$

which we first have to derive, and in particular enquire how the Planck and Boltzmann constants (h and k) arise, and what is meant by the free enthalpy, $\Delta G^{\ddagger} = \Delta H^{\ddagger} - T\Delta S^{\ddagger}$, of the "activated complex." The subscript "1" to k_1 indicates that this is a particular case of the more general k_n rate coefficient, the n denoting the "order of the reaction". The speed or rate of a chemical reaction is then k_n multiplied by (powers of) the concentrations or activities of the various reacting species. This means that the dimensions of k_n are not trivial, but in our case, k_1 has units of s^{-1}. The order of a reaction is generally to be determined by experiment.

13.2 First-Order Rate Equation

Imagine I have one gram of metal which was at one time pure artificial radioactive cobalt-60. Its atoms are decaying to become nickel atoms by each emitting a gamma photon, an electron and and an electron anti-neutrino:

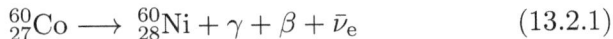

$$^{60}_{27}\text{Co} \longrightarrow {}^{60}_{28}\text{Ni} + \gamma + \beta + \bar{\nu}_e \qquad (13.2.1)$$

The decay is a quantum mechanical, stochastic process and so the probability that any particular atom decays is governed by Poisson statistics. At some particular time, t, the fraction of metal that is cobalt-60 is, say, x (the remaining fraction being nickel is $1 - x$).

---

[1]But the reader unfamiliar with these topics may skip without loss of continuity from the end of Section 13.3 to the beginning of Chapter 14.

The positive *rate of decay*, the rate at which the amount of cobalt-60 is depleting, is $-\dot{x}$ and this depends only on the amount of cobalt-60 remaining in the sample. (This is the principle on which carbon dating is based.) The rate equation is thereby

$$\frac{dx}{dt} = -k_1 x \qquad (13.2.2)$$

The solution to this equation is

$$x = x_0 e^{-k_1 t}$$

where x_0 is the fraction of cobalt-60 at time $t = 0$. The rate constant, k_1, has units of s^{-1}. If at $t = 0$ we have pure cobalt-60, then after the half life, $t_{\frac{1}{2}}$, has elapsed, we have $x = \frac{1}{2}$, and so the half life (5.3 years in this case) is related to the rate constant through $t_{\frac{1}{2}} = -\ln\frac{1}{2}/k_1 = 0.69/k_1$. Hence $k_1 = 4.1 \times 10^{-9}$ s^{-1}.

In chemistry, (13.2.2) is called a "first-order rate equation" because the "rate of reaction" depends linearly on the concentration, x. (I am a bit lax in using concentration and atom fraction interchangeably.) Radioactive decay is about the only reaction I can think of in which the rate is easy to understand. In general, chemical reactions involve any number of reactants and the rate is found experimentally usually to depend on integer powers of the concentrations of the reactants. For example, in the reaction (9.1.1), the rate may depend on the n_A^{th} power of c_A times the n_B^{th} power of c_B. Rates of reaction are of course of central importance and come under the very difficult discipline of *kinetics* in physical chemistry. Luckily for us, in corrosion, we only need to think about reactions like (13.2.1) of the type

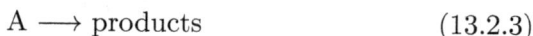

$$\text{A} \longrightarrow \text{products} \qquad (13.2.3)$$

which we can take to have first-order kinetics. On the other hand, we have deep difficulties of a pictorial or conceptual type because the detachment of a metal atom from the surface of an electrode and its journey as a positive cation through the "double layer" and into the bulk of the electrolyte are hard to conceive of within the conventional picture of "molecular dynamics", that is to say, the random collision of molecules and the generation of "activated states" by drawing energy from the heat bath.

There are at least three ways to "derive" Equation (13.1.1) and I will try and outline these in the following sections. Let us start with a very appealing picture due to Bockris and Reddy.

13.3 Rate Constant According to Bockris and Reddy

We plunge straight in and accept that the process whose rate we wish to calculate is the transfer of an ion from electrode to electrolyte. Bockris considers an idealisation of this as represented in Figure 13.1. We imagine two "boxes" separated by a distance, d, and the "reaction" is the process in which an ion, initially in box 1, transfers across the gap into box 2. We will call d_1 the distance from the ion to box 1 and $d_2 = d - d_1$ its distance from box 2. At what point do we say that "a jump has occurred" and the ion is on its way to box 2? If the process is completely symmetric, then obviously that is the point at which the ion is half way across (and travelling at some speed, v, in the direction of box 2 — more on the speed later). If the potential energy is not symmetric, say the energy of the ion

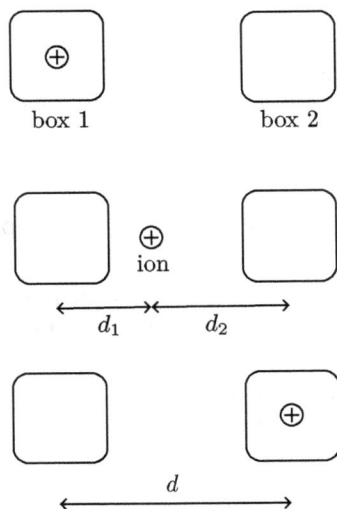

Fig. 13.1.   A charged species travels from the box on the left to the box on the right. In between the two, it is a distance d_1 from box 1. The separation between the boxes is d.

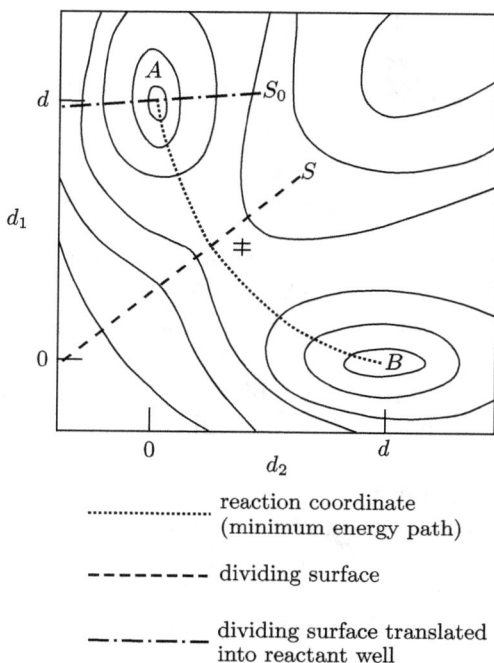

Fig. 13.2.   A two-dimensional potential energy surface represented by contours. During a chemical reaction, the "reactants" occupy the energy minimum labelled A; the "product basin" is labelled B. The path of lowest maximum energy is called the "reaction path" and at its maximum the system is at a saddle point in the energy landscape, denoted with the symbol $\ddagger$. The two broken lines represent dividing surfaces, see Section 13.7.

in box 2 is less than that in box 1, then the "point of no return" is not midway between the two. This is captured in a number called the "symmetry factor" which we will find plays a major role in the Butler–Volmer equation. Suppose that we know the potential energy of the system as a function of the coordinates d_1 and d_2. A contour map of this two-dimensional function is illustrated in Figure 13.2.

The path that the ion takes in the configuration space $\{d_1, d_2\}$ is called the "reaction coordinate". We take it that the path chosen is the one which entails the least amount of potential energy increase, namely the path that takes the ion over the saddle point in the energy landscape. We can then plot the potential energy as a function of the reaction coordinate in a plot such as Figure 13.3. In general, there

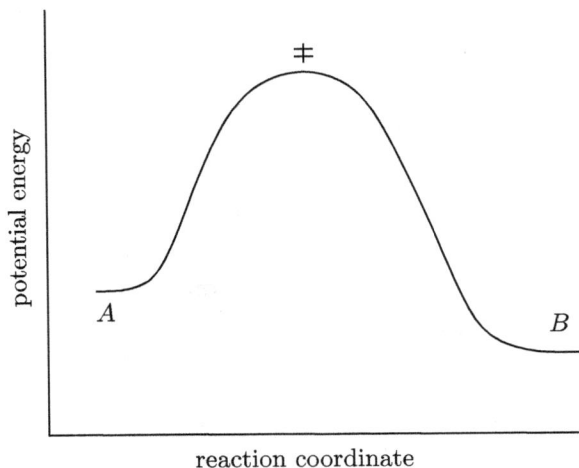

Fig. 13.3. A simple representation of the potential energy as a function of "reaction coordinate".

are not just two coordinates involved in such a process and so we are invited to try and imagine that a plot such as Figure 13.2 is somehow representative of an energy landscape in a multicoordinate configuration space; if there are N atoms involved in some thermally activated process, then we have to imagine a path in a $3N$-dimensional space. What characterises the saddle point is that there all the curvatures of the potential energy are positive *except for one*, namely the curvature along the direction of the reaction coordinate.[2] At that point, we talk of the existence of an "activated complex" always indicated by the symbol $\ddagger$. The maximum potential energy, which is encountered at the saddle point in the $3N$-dimensional space, is called the critical activation enthalpy, $\Delta H^{\ddagger}$; if there is an entropy change involved in the change from the "reactant basin" to the saddle point, then we talk of the free enthalpy of activation, $\Delta G^{\ddagger} = \Delta H^{\ddagger} - T\Delta S^{\ddagger}$. We will return to the complication of many coordinates later; for now, let us pursue Bockris's argument for the simple two-coordinate idealisation.

---

[2]In the simple example here, there are only two curvatures at the saddle point — one positive in sign and one negative, but we will generalise this picture in Section 13.7.

What characterises the state of the system at the saddle point? We think of this as a decisive situation: if the ion reaches the saddle point moving at a finite speed, v, then the reaction is bound to go to completion. The system will "fall down the energy hill". What is the unique indicator of this "point of no return"? It is the loosening of the hold that box 1 has on the ion: the bond connecting the two breaks. One way to describe that is to say that vibration (or stiffness) of the bond vanishes and is replaced by a translational degree of freedom. Bockris argues that this condition is reflected in an equality of the vibrational energy $h\nu$ and the translational energy kT. Hence at the moment of the reaction happening, there exists an activated state whose frequency of vibration in the direction of the reaction coordinate is $\nu = kT/h$ (6×10^{12} s^{-1} at 300K), where h is the Planck constant. His argument now is that the rate of transfer of the ion from box 1 to box 2 is the concentration of activated complexes, $c^{\ddagger}$, times the number of times per second on average the complex makes "an attempt" to fall down the hill (from that language, the name "attempt frequency," ν, has been coined). This is a central postulate of the transition state theory and sometimes it is formulated in terms of the speed, v, rather than the frequency, ν. So we have clearly moved from the notion of a single transfer event to the consideration of an ensemble of systems since we are enquiring into the concentration of activated complexes. Now we come to a second postulate of transition state theory and one which on the face of it is hard to take seriously. This is that we allow that in the ensemble there exists *an equilibrium between the reactants and the activated complexes*. For the case of reaction (13.2.3), this implies the equilibrium

$$A \rightleftharpoons A^{\ddagger}$$

Following the arguments in Section 9.1, there must be an equilibrium constant (let's use concentration for now)

$$K^{\ddagger} = \frac{c^{\ddagger}}{c} \qquad (13.3.1)$$

and the molar standard free enthalpy of activation must be

$$\Delta G^{\ddagger} = -RT \ln K^{\ddagger} \qquad (13.3.2)$$

This is quite an astonishing assertion in view of the thought that the activated complex is extremely short lived and has at least one imaginary frequency[3] of vibration — although in Bockris's argument we actually have $\nu = kT/h$ real and positive. Nonetheless, we are assured by the textbooks that this is by no means a weak postulate of transition state theory.

Finally we can complete the derivation. We are looking for the rate of reaction in the form of (13.2.2), namely

$$\dot{c} = -k_1 c$$

and if we take unit concentration of reactants, we find that the *rate of reaction* is, in moles m^{-3} s^{-1},

$$r = k_1 c = k_1 \times \textbf{one}$$

and hence,

$$k_1 = r = \text{attempt frequency} \times \text{conc}^{n\cdot} = \nu c^{\ddagger} = \frac{kT}{h} e^{-\Delta G^{\ddagger}/RT}$$

$$(13.3.3)$$

as required in (13.1.1). In preparing this text, I may have become a bit obsessed with the various ways in which the formula for the rate coefficient may be derived, so we will now go into a rather lengthy and possibly over detailed excursus. The reader who is content with Bockris's argument as it stands may profitably skip now to Chapter 14.

13.4 Velocity at the Saddle Point

Sometimes the rate coefficient is couched in terms of the velocity, v, introduced in the last section, rather than the frequency, ν. Using Figure 13.2 and some results from statistical mechanics, we can develop a formula for this speed. Consider that the reaction path shown in Figure 13.2 is parametrised by a coordinate x as $x(d_1, d_2)$. The ion in motion has a mass m and a speed $\dot{x}$ and so its kinetic energy is $\frac{1}{2} m \dot{x}^2$. If the speeds of an ensemble of these particles (ions in the example of Section 13.3) have an equilibrium distribution,

---

[3] $\nu^2 < 0$ at the top of the potential energy curve in Figure 13.3.

then the average speed is given by a ratio of configuration integrals, namely

$$\langle v \rangle = \frac{\int_0^\infty e^{-\frac{1}{2}\beta m \dot{x}^2}\, \dot{x}\, \mathrm{d}\dot{x}}{\int_{-\infty}^\infty e^{-\frac{1}{2}\beta m \dot{x}^2}\, \mathrm{d}\dot{x}} \qquad (13.4.1)$$

in which we use the standard notation, $\beta = 1/kT$. Note that the lower limit in the numerator is zero and not $-\infty$ because we are only interested in states in which the activated complex is flying in the positive x direction and hence the reaction is occurring. We need these two standard integrals:

$$\int_{-\infty}^\infty e^{-ax^2}\, \mathrm{d}x = \sqrt{\frac{\pi}{a}} \qquad (13.4.2a)$$

$$\int_0^\infty x\, e^{-ax^2}\, \mathrm{d}x = \frac{1}{2a} \qquad (13.4.2b)$$

There results

$$\langle v \rangle = \sqrt{\frac{1}{2\pi\beta m}} = \sqrt{\frac{kT}{2\pi m}} \qquad (13.4.3)$$

Note that since the axes of Figure 13.2 are distance, the units of v are indeed m s^{-1}. However, it is not necessary to require that v is actually the speed of some particle or ion. Indeed, we will generalise this picture in Section 13.7. If the reaction is dominated by the passage of a single particle in the presence of other particles which are essentially fixed in position, then we may take it that m really is the mass of the particle. An example would be the diffusion of atomic hydrogen through a lattice of metal atoms. Otherwise, and indeed usually, we must interpret m as an effective mass which characterises the reaction.

13.5 Equilibrium Constant in Terms of Partition Function

We consider the equilibrium

$$A \rightleftharpoons B \qquad (13.5.1)$$

and imagine that the reactants and products are simple systems such as molecules having a set of fixed single particle energy levels as

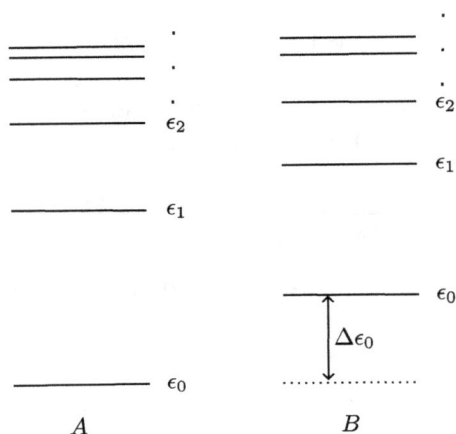

Fig. 13.4.   A set of discrete quantum mechanical energy levels of two entities, A and B. Note that the lowest energy levels of each do not necessarily coincide, the difference being denoted $\Delta\epsilon_0$.

sketched in Figure 13.4. Each then has a microcanonical partition function

$$z'_A = \sum_{i=0}^{\infty} e^{-\epsilon_i(A)/kT}$$

and

$$z'_B = \sum_{i=0}^{\infty} e^{-\epsilon_i(B)/kT}$$

Each of these is relative to its own lowest eigenvalue ϵ_0. We need to choose a common energy zero so we use $\epsilon_0(A) = 0$. When each partition function is expressed relative to this common zero of energy, we get

$$z_A = z'_A$$

and

$$z_B = \sum_{i=0}^{\infty} e^{-(\epsilon_i(B)+\Delta\epsilon_0)/kT}$$
$$= z'_B e^{-\Delta\epsilon_0/kT}$$

The absence of a prime indicates partition functions referred to a common energy zero. z_B is the sum of all probabilities that the equilibrium (13.5.1) finds itself in one of the energy levels of B and similarly for A. It follows that the concentrations of B and A are in the same ratio as their partition functions:

$$\frac{c_B}{c_A} = \frac{z_B}{z_A} = \frac{z_B'}{z_A'} e^{-\Delta\epsilon_0/kT} \tag{13.5.2}$$

Indeed, this result is very general and not just confined to single particle partition functions. If the reactants or products are made of an assembly of N identical particles, then their canonical partition function is

$$Z = \frac{z^N}{N!}$$

and it becomes axiomatic that if both partition functions are referred to the same energy zero:

$$\frac{c_B}{c_A} = \frac{Z_B}{Z_A} \tag{13.5.3}$$

We should note for future reference that the single particle partition function can be approximately factored into its translational, rotational, vibrational and electronic contributions

$$z' = z_{tr}\, z_{rot}\, z_{vib}\, z_{el}$$

(omitting primes on the right-hand side) well-known formulas being

$$z_{vib} = \left(1 - e^{-h\nu/kT}\right)^{-1} \tag{13.5.4}$$

$$z_{tr} = \frac{V}{\Lambda^3}$$

where

$$\Lambda = \frac{h}{\sqrt{2\pi mkT}}$$

is the thermal de Broglie wavelength and m is the mass of the particle. V is volume.

The (Helmholtz) free energy is given by a central formula of statistical mechanics:

$$F = -kT \ln Z$$

Let us use a series of approximations to find a relation between the single particle partition functions and the equilibrium constant for the reaction (13.5.1). We proceed as follows. For one mole (L particles) of a substance,

$$F = -kT \ln Z \tag{13.5.5}$$

$$= -kT \ln \left(\frac{z^L}{L!} \right) \tag{13.5.6}$$

$$= -LkT \ln z + kTL \left(\ln L - 1 \right)$$

$$= -RT \ln \frac{z}{L}$$

in which L is the Avogadro constant: the number of particles per mole. I have used the slacker Stirling approximation, $\ln N! = N(\ln N - 1)$ and neglected one compared to $\ln L = 55$, and used the relation between the Boltzmann and gas constants, $R = Lk$. Finally, observe that in condensed matter we may ignore the difference between free energy and free enthalpy, $G = F + pV$, since bulk properties of solids and liquids are virtually unchanged between ambient and near-zero pressure. We can thereby assert that the standard state free enthalpy per mole in condensed matter is[4]

$$G^\circ = -RT \ln \frac{z^\circ}{L}$$

as long as we may factor the partition function into identical single particle partition functions as in (13.5.6). In view of this, I can write

---

[4]This result is exact for an ideal gas as you can see by adding $pV = RT$ to (13.5.5). Then I don't need to neglect the one.

the standard free enthalpy change of reaction (13.5.1) as

$$\Delta G^\circ = -RT \ln \left(\frac{z_B^\circ}{z_A^\circ} e^{-\Delta \epsilon_0 / kT} \right) = -RT \ln K$$

defining standard state partition functions referred to a common energy zero. This identifies an equilibrium constant for the reaction (13.5.1) as the ratio of the probability of finding the equilibrium system in one of the states of the reactants to the probability of finding the system in one of the states of the products. We see from (13.5.2) that

$$K = \frac{c_B}{c_A}$$

when reactant and product are in a standard state, which is indeed consistent with our definition of equilibrium constant in Section 9.1 if we assume that activity coefficients are one, that is, that the system is in some sense "ideal."

This observation throws up a serious disconnect in the relation between equilibrium constant and partition function. Suppose we return to the reaction of Section 9.1:

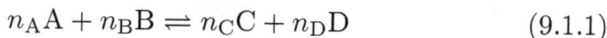

$$n_A A + n_B B \rightleftharpoons n_C C + n_D D \tag{9.1.1}$$

Its equilibrium constant is

$$K = \frac{a_C^{n_C} a_D^{n_D}}{a_A^{n_A} a_B^{n_B}} \tag{9.1.3}$$

(Clearly, this can be extended to any number of reactants and products.) On the other hand, the ratio of partition functions gives us concentrations since these relate directly to the probabilities of the equilibrium being found in one of the microcanonical states. Hence we must write

$$\frac{c_C^{n_C} c_D^{n_D}}{c_A^{n_A} c_B^{n_B}} = \frac{(z_C/V)^{n_C} (z_D/V)^{n_D}}{(z_A/V)^{n_A} (z_B/V)^{n_B}}$$

This means that, in our simple "unimolecular" reaction (13.5.1), we should write the relation between equilibrium constant and partition function as

$$K = \frac{\gamma_B c_B}{\gamma_A c_A} = \frac{\gamma_B z_B^\circ}{\gamma_A z_A^\circ}$$

In the following section when we apply this formula to the activated complex, you will see how unsatisfactory this is. Not only do we need to imagine that we can apply equilibrium thermodynamics to a mechanically unstable object, we also need to assert that we can somehow define an "ideal" state and furthermore to give a value to its activity coefficient. Some textbooks avoid this by defining equilibrium constant directly in terms of concentration, or in terms of partial pressure, but I'm not sure whether that doesn't just sweep the problem under the carpet.

Ultimately, the whole of rate theory rests on *experimental* measurements that show that the rate of a reaction is in proportion to some powers of the concentrations of the reactants. If the observations had shown a closer correlation with activities rather than concentrations, then maybe the theory would have looked a little different.

13.6 Rate Coefficient in Terms of Partition Function

If we apply reasoning from the previous section to the equilibrium,

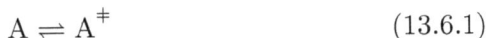

$$A \rightleftharpoons A^{\ddagger} \tag{13.6.1}$$

between an initial state and the activated complex as in Section 13.3, then we may write (13.5.2) in the form

$$K^{\ddagger} = \frac{z^{\ddagger}}{z} e^{-\Delta U/kT}$$

in which z stands for z_A and ΔU is the difference in energy between the lowest energy levels of the activated complex and the initial state: the equivalent of $\Delta \epsilon_0$ in Figure 13.4. In this way, ΔU is the zero temperature activation energy (in the sense that at $T = 0$ the species are all in their ground states). Now, by definition, the rate of change of the amount of reactant in the initial state of the system is, in comparison to (13.3.3),

$$-\dot{c} = k_1 c = \text{conc}^n \text{ of activated complexes times}$$

their frequency of crossing the saddle point

$$= c^{\ddagger} \nu^{\ddagger}$$

Now from (13.5.2),

$$c^{\ddagger} = c\, \frac{z^{\ddagger}}{z}\, e^{-\Delta U/kT}$$

so it follows that

$$k_1 = \nu^{\ddagger} \frac{z^{\ddagger}}{z}\, e^{-\Delta U/kT}$$

We factorise out from $z^{\ddagger}$ the contribution from the vibrational mode whose oscillations are in the direction of the reaction coordinate at the saddle point:

$$z^{\ddagger} = z^{\ddagger}_{\mathrm{vib}} z^{\ddagger *}$$

(The star reminds us that this partition function has one fewer degree of freedom.) We recall the argument of Bockris in Section 13.3 that this vibrational mode is a borderline translation so its frequency must be tending to zero. Therefore, from (13.5.4), we have

$$z^{\ddagger}_{\mathrm{vib}} = (1 - e^{-h\nu^{\ddagger}/kT})^{-1} = \frac{kT}{h\nu^{\ddagger}} \quad \left(h\nu^{\ddagger} \ll kT\right)$$

The result is

$$k_1 = \frac{kT}{h} \frac{z^{\ddagger *}}{z}\, e^{-\Delta U/kT} \tag{13.6.2}$$

It is very typical, as we shall see, that the rate coefficient involves a ratio of partition functions in the numerator of which there is the partition function of the activated complex *with one degree of freedom removed*. This is handy because since we are at a saddle point, there is one normal mode whose frequency is imaginary, so it's well that we are able to factor it out. Please note that the reduction by one degree of freedom as indicated by the asterisk, in this case, does not entail a change in dimension since the partition functions here are dimensionless. We see in Section 13.7 that in a more general case, the reduction in degrees of freedom also results in a change in the dimension of a partition function as this has an impact on the interpretation of a formula equivalent to (13.6.2).

Finally, we can eliminate the partition functions and replace the Arrhenius activation energy with the activation free enthalpy in the following way. We write the van 't Hoff isotherm for the reaction (13.6.1)

$$\Delta G^{\circ \ddagger} = -RT \ln K^{\ddagger}$$

and insert this into

$$k_1 = \frac{kT}{h} K^{\ddagger}$$

and we get the expression for the rate coefficient in its familiar form (familiar that is to physical chemists):

$$k_1 = \frac{kT}{h} e^{-\Delta G^{\circ \ddagger}/RT} \tag{13.6.3a}$$

$$= \frac{kT}{h} e^{-\Delta S^{\circ \ddagger}/R} e^{-\Delta H^{\circ \ddagger}/RT} \tag{13.6.3b}$$

$\Delta G^{\circ \ddagger}$, $\Delta H^{\circ \ddagger}$ and $\Delta S^{\circ \ddagger}$ are respectively the free enthalpy (Gibbs free energy), enthalpy and entropy of activation in a standard state. There is a glaring inconsistency in this last argument because it ignores the fact that we have already pulled out the imaginary vibration frequency. However, Walter Moore insists that this "...has no appreciable effect on the practical applications of the equation." Given the messy nature of the problem, the difficulty in finding a pictorial representation of the process and the obscurity in defining what is an "activated complex" it is probably not surprising that there are logical difficulties. However, if you ask is there any rigorous way to write down the rate coefficient, then read on to the following section.

13.7 Formulation from First Principles

If you are in despair that a plausible explanation of the rate coefficient has so far eluded us, then join me as we turn to solid state physics for a more satisfactory development. This is associated with the names of C. A. Wert and the legendary Clarence Zener of Chicago and established in the seminal paper by George H. Vineyard (*J. Phys. Chem. Solids*, **3**, 121 (1957)). Return to Figure 13.2 but now allow it to represent a case where $N/3$ atoms are interacting. The following development was originally made for the diffusion of an atom in a metal, so the point marked "A" may represent the state in which there is a vacancy next to an atom which is about to jump into the vacant site surmounting a barrier in the energy landscape. The point marked "B" may represent the crystal after the atom has jumped into the vacant site, but in general the points A and B need not be symmetrically disposed and there may be an energy change associated with the reaction A $\rightarrow$ B. Each atom has three position coordinates, denoted $x_1, x_2, \ldots x_N$, so Figure 13.2 must now be interpreted as a set of contours in an N-dimensional, not two-dimensional, space. The broken line is constructed to intersect the saddle point and to be perpendicular to all the contours. It is called a *dividing surface* and is a hyperplane of $N - 1$ dimensions. For the system to move from the basin A into the basin B through the dividing surface, there must be some "flow of probability" which we try and calculate. Associated with each position coordinate, x_i, there is a velocity $\dot{x}_i$. In what follows, we simplify matters and insist that all atoms have the same mass, m. If this were not the case, then we would rescale all the coordinates (as in the original work) to $y_i = \sqrt{m_i}x_i$.

Because we have migrated from physical chemistry to solid state physics (and who can judge which is closer to the detachment of an ion from a metal into an electrolyte?), we dispense with the rather artificial notion of "concentration", and indeed we did this in the argument given by Bockris in Section 13.3 by setting $c = 1$. Instead, we ask a completely well posed question: what is the average lifetime, τ, of all configurations represented by that part of the configuration space to one side of the dividing surface, namely the side containing configuration A? Then the rate of transition from A to B is

$$k_1 = \frac{1}{\tau} = \frac{\mathcal{I}}{Z_A} \qquad (13.7.1)$$

Here, Z_A is the number of points in the configuration space on the "A-side" of the dividing surface and $\mathcal{I}$ is the number of such points crossing the dividing surface in the direction of B per unit of time. Z_A is a partition function that I write to within a multiplicative normalisation that will cancel in the end so I'll leave it out:

$$Z_A = \int \ldots \int_A e^{-\beta U} \mathrm{d}x_1 \mathrm{d}x_2 \ldots \mathrm{d}x_N$$

where $U = U(x_1, x_2 \ldots x_N)$ is the potential energy surface whose contours in the N-dimensional space are sketched in Figure 13.2.

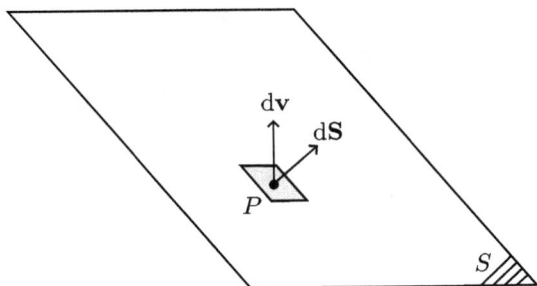

Fig. 13.5. Two-dimensional representation of a small patch of $N-1$-dimensional potential energy surface. At the point, P, on the surface, an infinitesimal area has a normal vector denoted d**S** and the reaction velocity at that point is **v**.

It is worthwhile to point out what are the *dimensions* of Z_A in the absence of a normalising constant. This is an integral over configuration space (restricted to that part which is on the reactant side of the dividing surface) and so has the dimensions of $[\text{length}]^N$. Now we need to calculate $\mathcal{I}$. Figure 13.5 shows a patch of the dividing surface (in $N-1$ dimensions!), a point, P, and an element of surface at P whose normal is the vector d**S**. If **v** is the velocity of points in configuration space (see Section 13.4), then we will have that the infinitesimal element of current crossing d**S** is

$$d\mathcal{I} = d\mathbf{S} \cdot \langle \mathbf{v} \rangle$$

We need to know what the statistical distribution of the velocity is at P. The number of configuration space points at P having a speed v in the range $dv = d\dot{x}_1 d\dot{x}_2 \dots d\dot{x}_N$ is

$$\rho(P,v)dv = \frac{e^{-\beta U(P)} e^{-\frac{1}{2}\beta m v^2} \, d\dot{x}_1 d\dot{x}_2 \dots d\dot{x}_N}{\int_{-\infty}^{\infty} \dots \int_{-\infty}^{\infty} d\dot{x}_1 d\dot{x}_2 \dots d\dot{x}_N \, e^{-\frac{1}{2}\beta m \dot{x}_1^2} e^{-\frac{1}{2}\beta m \dot{x}_2^2} \dots e^{-\frac{1}{2}\beta m \dot{x}_N^2}}$$

The denominator is a product of identical terms as in (13.4.1) so using (13.4.2a), we get

$$\text{denominator} = \left(\sqrt{\frac{2\pi}{\beta m}} \right)^N = \left(\frac{2\pi kT}{m} \right)^{\frac{1}{2}N}$$

We now need to find

$$d\mathcal{I} = d\mathbf{S} \cdot \int \mathbf{v} \, \rho(P,v) \, dv$$

and the domain of integration is to be such that the flow is all in the direction A→B, that is, $d\mathbf{S} \cdot \mathbf{v} > 0$. A neat way to achieve this is to imagine that we have rotated the coordinate system so that one of the $\dot{x}_i$, say $\dot{x}_1$, is directed parallel to $d\mathbf{S}$. This means that all other velocities are lying in the dividing plane so they factor out. We get

$$d\mathcal{I} = \left(\frac{2\pi kT}{m}\right)^{-\frac{1}{2}N} e^{-\beta U(P)} \, dS_1 \int_0^\infty \dot{x}_1 \, e^{-\frac{1}{2}\beta m \dot{x}_1^2} \, d\dot{x}_1$$

$$\times \int_{-\infty}^\infty e^{-\frac{1}{2}\beta m \dot{x}_2^2} \, d\dot{x}_2 \int_{-\infty}^\infty e^{-\frac{1}{2}\beta m \dot{x}_3^2} \, d\dot{x}_3 \ldots \int_{-\infty}^\infty e^{-\frac{1}{2}\beta m \dot{x}_N^2} \, d\dot{x}_N$$

As in (13.4.1), we only integrate over the velocities in the forward direction across the dividing surface and this first integral evaluates to kT/m (13.4.2a). The second line here is a product of $N-1$ identical integrals exactly one fewer than in the denominator we evaluated just earlier. So these all but cancel the prefactor:

$$\left(\frac{2\pi kT}{m}\right)^{-\frac{1}{2}N} \left(\frac{2\pi kT}{m}\right)^{\frac{1}{2}(N-1)} = \sqrt{\frac{m}{2\pi kT}}$$

When this is multiplied by the factor kT/m from the integral over $\dot{x}_1$, we get a prefactor $\sqrt{kT/2\pi m}$. Now we obtain the total current by integrating $d\mathcal{I}$ over the dividing surface:

$$\mathcal{I} = \sqrt{\frac{kT}{2\pi m}} \int \ldots \int_S e^{-U/kT} \, dS$$

Now we combine this with (13.7.1) and obtain our final result for the rate coefficient:

$$k_1 = \sqrt{\frac{kT}{2\pi m}} \frac{\int \ldots \int_S e^{-U/kT} \, dS}{\int \ldots \int_A e^{-U/kT} \, dV} = \sqrt{\frac{kT}{2\pi m}} \frac{Z_S^*}{Z_A} \tag{13.7.2}$$

The prefactor, $\sqrt{kT/2\pi m}$, is the velocity we calculated in (13.4.3). Again, this has the actual units of velocity, m s^{-1}, as long as the energy surface is constructed in direct space (not, for example, scaled by $\sqrt{m_i}$ as in the original work by Vineyard). However, one could imagine that in a chemical reaction the reaction coordinate does not have dimensions of length as measured in metre, in which case we may think of a generalised "velocity" along the reaction coordinate, possibly having unusual "length" dimensions. At any event, the expression (13.7.2) has the expected units of s^{-1}. This is because the velocity prefactor has dimensions of ["length"][time]$^{-1}$ while the ratio of partition functions has dimensions of ["length"]$^{-1}$ since the numerator has one fewer degree of freedom and hence one fewer factor of dx than the denominator.

Equation (13.7.2) is reminiscent of our earlier formula (13.6.2), namely a ratio of two partition functions, this time multiplied by the velocity rather than by a frequency. It is very important to note that as in (13.6.2) the partition function in the numerator has one degree of freedom fewer than that in the denominator. I have indicated this again with an asterisk.

The formula (13.7.2) does not, on the other hand, have the familiar Arrhenius form involving the exponential of an activation energy, or free enthalpy, over kT. Nevertheless, we can cast it into both such forms.

Firstly, if we use the approximation of small vibrations, then in the manner of the theory of lattice vibrations, we expand the potential energy U both at the configuration point A at the minimum of energy in the "A-basin" and again at the point A‡ where the reaction coordinate intersects the dividing surface:

$$U \approx U(A) + \frac{1}{2} \sum_{i=1}^{N} m\omega_i^2 q_i^2 \qquad \text{about A}$$

$$U \approx U(A^{\ddagger}) + \frac{1}{2} \sum_{i=1}^{N-1} m\omega_i^{\ddagger 2} q_i^{\ddagger 2} \qquad \text{about A}^{\ddagger}$$

Here, the ω_i and $\omega_i^{\ddagger}$ are normal mode angular frequencies ($\omega = 2\pi\nu$) and the q_i and $q_i^{\ddagger}$ are associated normal mode coordinates. In the second expansion, the vibrations are constrained to remain in the dividing surface so there is one fewer mode, namely that mode is missing that would correspond to vibration in the direction of the reaction coordinate and which would hence possess an imaginary frequency. In this harmonic approximation, the partition functions can be evaluated.[5] They are

$$Z_S^* = \prod_{i=1}^{N-1} \sqrt{\frac{2\pi kT}{m\omega_i^{\ddagger 2}}} e^{-U(A^{\ddagger})/kT}$$

$$Z_A = \prod_{i=1}^{N} \sqrt{\frac{2\pi kT}{m\omega_i^2}} e^{-U(A)/kT}$$

Putting this into (13.7.2), there is cancellation of some 2πs and ms and the result is

$$k_1 = \nu' e^{-(U(A^{\ddagger}) - U(A))/kT} \tag{13.7.3}$$

where

$$\nu' = \frac{\nu_1 \nu_2 \ldots \nu_N}{\nu^{\ddagger}_1 \nu^{\ddagger}_2 \ldots \nu^{\ddagger}_{N-1}} \tag{13.7.4}$$

This is in the form of (13.6.2) in the sense that the Arrhenius term is the exponential of a *potential* energy, or zero temperature enthalpy.

Secondly, to cast the rate coefficient in terms of an exponential of a free enthalpy of activation, so that we can finally make contact with the putative formula (13.1.1), we return to (13.7.2) and Figure 13.2. In addition to the dividing surface, S, we imagine a hypersurface of the same reduced dimension, S_0, as sketched in Figure 13.2, which is the dividing surface *translated* so that it now intersects the minimum of the reactant basin at A. There are now two reduced dimension partition functions:

$$Z_S^* = \int \ldots \int_S e^{-U/kT} \, dS$$

$$Z_0^* = \int \ldots \int_{S_0} e^{-U/kT} \, dS$$

Now take Equation (13.7.2) and multiply and divide by Z_0^*:

$$k_1 = \sqrt{\frac{kT}{2\pi m}} \frac{Z_0^*}{Z_A} \frac{Z_S^*}{Z_0^*}$$

---

[5]Put the expansion of U into $\int \ldots \int e^{-U/kT} dq_1 dq_2 \ldots$ and use (13.4.2a).

Now comes the *coup de grâce*. Because the free energy is $F = -kT \ln Z$, we have

$$\frac{Z_S^*}{Z_0^*} = e^{-\Delta F/kT}$$

and we interpret ΔF as the reversible work done in taking the configuration space points which are constrained to belong to S_0 into those points that are constrained to belong to S. That is to say, we carry the system from the reactant basin to the saddle point. In this sense, $\Delta F = \Delta U - T\Delta S$ is the free energy of activation, and ΔU and ΔS are potential energy and entropy of activation, exactly as in (13.1.1) allowing that we need not distinguish enthalpy and internal energy in condensed matter (p = ambient and $p \approx 0$ lead to little difference in bulk properties). Finally, then we have for the rate coefficient,

$$k_1 = \sqrt{\frac{kT}{2\pi m}} \frac{Z_0^*}{Z_A} e^{-\Delta F/kT} = \nu\, e^{-\Delta F/kT} \tag{13.7.5}$$

Now the ratio is of partition functions both belonging to the reactant basin: the denominator being the number of all configuration space points to the reactant side of the dividing surface and the numerator being the same but constrained to that part of the space that excludes those configurations which are in some sense "parallel" to the dividing surface as indicated in Figure 13.2. It is most significant that in this formulation of the rate coefficient no reference is made to the properties of the activated complex in contrast to (13.7.2) which requires us to know the partition function at the dividing surface. Instead, in (13.7.5), the nature of the activated complex is relegated to an activation free energy and only partition functions evaluated at the reactant state are called for. (This is particularly evident in quantum transition state theory in which the activated state becomes almost irrelevant since an atom may anyway *tunnel* through the barrier.) I should point out that the choice of S_0 is not unique — but then neither is the choice of dividing surface if the energy landscape is not symmetrical. In the small vibration approximation, it is possible as in (13.7.3) to identify the prefactor, ν, in (13.7.5) with a ratio of frequencies similar to (13.7.4). But this time there is no reference to the vibrations of the activated complex since the frequencies belonging to A and to S_0 are those of the reactant basin. The frequency prefactor, ν, is related to the frequency ν' in (13.7.3) via

$$\nu = \nu' e^{-\Delta S/k}$$

Then, in the small vibration approximation, if you use the result in statistical mechanics that $S = k \ln Z + U/T + \text{const.}$, you find in contrast to (13.7.4)

$$\nu = \frac{\nu_1 \nu_2 \ldots \nu_N}{\nu_1^0 \nu_2^0 \ldots \nu_{N-1}^0} \tag{13.7.6}$$

I can only see one way in which this ratio can be cast into a simple kT/h as required by (13.1.1). This is if we assume all these frequencies, which belong to the reactant basin A, are single Einstein frequencies all but one of which cancel in the ratio since there is one more frequency factor in the numerator that will not cancel. If this mode is fully excited, then it could be expressed as kT/h, but the contrast with the earlier argument leading to (13.6.2) is that in that case we identified kT/h with a mode of the activated complex, not of the reactants. For this reason, the formulation leading to (13.7.5) is appealing because it does not deal in unmeasurable, unthermodynamic quantities, such as frequencies of, or equilibria with, fictitious "activated complexes."

We can now leave rate theory and return to electrochemistry, but the reader may be interested to know that whereas in the 1950s (13.7.5) was mostly a theoretical construction, nowadays with computational methods for calculation of partition functions, prediction of rate coefficients is feasible even when using reliable electronic structure-based total energy and force methods. Indeed, the theory can be extended by replacing the classical partition functions in (13.7.5) with quantum partition functions which can be calculated using Monte Carlo methods. This then allows a calculation of transition rates that include quantum tunnelling and zero point energy. An example of this in the context of hydrogen diffusion in iron can be found in A. T. Paxton and I. H. Katzarov, *Acta Materialia*, **103**, 71 (2016).

As a final comment, we have noted that one of the postulates of the classical transition state theory is that if the system finds itself at the saddle point with a finite velocity in the direction away from the reactant basin, then the reaction *will* occur. There is a caveat to this which arises from quantum mechanics, namely that there is a finite probability that even under this condition the atoms will draw back from the brink and return to the reactant state. This possibility is expressed in terms of a *transmission coefficient* which is a number less than one which multiplies the rate coefficient. Nowadays, transmission coefficients can also be calculated using molecular dynamics techniques.

Chapter 14

Single Electrode in Equilibrium and at an Overpotential

14.1 Detailed Balancing

We return to what is really a thought experiment of placing a coupon of pure metal into a solution of its ions. This establishes an equilibrium in the reaction

$$M^{n+}(aq) + ne^{-}(metal) \rightleftharpoons M(metal) \qquad (14.1.1)$$

which is a special case of Equation (9.6.1). Depending on the initial concentration of M^{n+} ions in the solution, this reaction will proceed to the left or to the right until metal ions in both phases come to equilibrium, that is, the electrochemical potentials of M^{n+} ions are equal in the electrolyte and in the metal.[1] As I have pointed out at the beginning of Chapter 13, even if the free enthalpy of oxidation dictates that this reaction should continue until all the metal has dissolved, the charging of the metal by the electrons liberated by oxidation of the metal prevents the backward reaction from proceeding; this is because of the electric self-energy associated with a charged object and indeed there is an additional self-energy due to charging of the electrolyte. In addition, the transfer of M^{n+} into the

---

[1]What we mean by the "electrochemical potential of ions in the metal" is understood within the decomposition of the chemical potential of atoms in the metal as $\tilde{\mu}_M = \tilde{\mu}_{M^{n+}} + n\tilde{\mu}_e$; admittedly, a strange way to think of metallic bonding.

solution becomes increasingly more costly as the metal's negative charge grows. In fact, it continues until the metal becomes charged to the extent of establishing a certain equilibrium electric potential difference across the interphase:

$$\Delta\phi_{eq} = \phi_m - \phi_s \qquad (14.1.2)$$

In principle, this single electrode potential cannot be measured. On the other hand, if the interphase were ideally polarisable and we connect it to an ideally reversible electrode such as a standard hydrogen electrode, then we could measure it. Indeed, if the electrolyte solution is at standard activity of M^{n+} ions, then the equilibrium potential may be read off from the electrochemical series Table 9.3, Section 9.5. If the concentration of M^{n+} ions is different from unit activity, then $\Delta\phi_{eq}$ can be found by application of the Nernst equation (9.6.2).

As a consequence of the equilibrium potential difference, $\Delta\phi_{eq}$, there will exist an electric field within the interphase, and the M^{n+} ion will have to do work against this electric field (the positive metal is repelling the ion and the negative electrolyte is attracting it back since it leaves behind a positive anion).

"Equilibrium" of this reaction does not mean that nothing is happening. What it does mean is that the forward and backward reactions are happening at the same rate (like the angels ascending and descending Jacob's ladder). This is called the *principle of detailed balancing* and is a direct consequence of the quantum and statistical mechanics (see Walter J. Moore in Further Reading, Appendix D for an explanation).

We shall examine the forward and backward rates of the reaction (14.1.1) separately, since unlike many activated processes both in chemistry and physics, there is a very large asymmetry when we compare the detachment of a metal atom from a surface to the migration of an ion from solution onto the surface of the metal. The former may well be better described by solid state kinetics and the latter by solution kinetics. You may have arrived here from Section 13.3 in which case we are agreed that the rate of some process is the product of a "concentration" or activity and a "rate coefficient" which may be written reasonably generally as

$$k_1 = \nu e^{-\Delta G^{\ddagger}/RT}$$

in which in the "frequency prefactor" is usually given in textbooks as

$$\nu = \frac{kT}{h}$$

or you may prefer a more rigorous form such as in Equation (13.7.6).

14.2 Electronation

The forward reaction of (14.1.1) is electronation (see Chapter 1). The M^{n+} ion moves from the outer Helmholtz plane (OHP) through the "double layer" and attaches to the metal surface, which then acquires a charge of $+ne$. Accompanying this process there is necessarily an electric current and if one mole of M^{n+} ions make the journey across a unit area of metal surface per second, then there arises an electronation current density, i_e, which we can find if we know the rate of the electronation reaction. If the concentration of M^{n+} ions is $\bar{c}_{M^{n+}}$ *moles per unit area* (not volume), then the rate of the reaction in moles per square metre per second is

$$r_e^0 = \nu_e \, \bar{c}_{M^{n+}} \, e^{-\Delta G_e^{\ddagger}/RT}$$

where $\Delta G_e^{\ddagger}$ is the free enthalpy of activation of electronation and the superscript, zero, denotes zero electric field. We might call $\Delta G_e^{\ddagger}$ the "chemical activation free enthalpy".

Indeed, this neglects the effect of the electric field. If the ion is moving against the field generated by the electric potential difference, $\Delta\phi_{eq}$, then this effectively *raises* the activation barrier. If the total potential difference between the OHP and the metal surface is $\Delta\phi_{eq}$, then the amount of work done against the electric field per mole of ions is *not* $nF\Delta\phi_{eq}$. This is because the ion only needs to do work against the electric field until it reaches the point somewhere between the OHP and the metal surface at which the free enthalpy of activation is at its maximum. After that the ion may traverse the remainder of the double layer without doing further electric work. This leads us to a central parameter in electrochemistry, the so-called *symmetry factor*, β. This is the *fraction of the total distance from the OHP travelled over which electrical work must be done by the ion.*

Hence the additional energy that needs to be supplied to the ion to complete the electronation reaction is $\beta nF\Delta\phi_{eq}$ and this essentially *lifts* the height of the free enthalpy barrier. The rate becomes

$$r_e = \nu_e \, \bar{c}_{M^{n+}} \, e^{-(\Delta G_e^{\ddagger} + \beta nF\Delta\phi_{eq})/RT}$$

$$= r_e^0 \, e^{-\beta nF\Delta\phi_{eq}/RT} \qquad (14.2.1)$$

It follows that the equilibrium electronation current density in A m^{-2} is

$$i_e^{eq} = r_e \, nF \qquad (14.2.2)$$

where F is the Faraday constant. We have made two principal approximations here. One is that the electric field is constant across the double layer so that the potential drop is linear. The second is that the only source of electric field is the surface charge on the metal and the neutralising surface charge density distributed over the outer Helmholtz plane of the electrolyte. Actually, even if the metal were neutral, there will be alignment of water molecules and ions in the solution reacting to image forces from the metal which may result in a dipole electric field in the double layer. Actually, we will see in a moment that all the numerous approximations we have made get swept up into a quantity which we call equilibrium exchange current density, which can anyway be measured and we won't have to worry any more where it comes from!

14.3   De-electronation

For the moment, we are considering the equilibrium state so that simultaneously there is a current in the opposite direction corresponding to de-electronation — the backward reaction of (14.1.1). We should write something like

$$r_d^0 = \nu_d \, (\text{areal concentration}) \, e^{-\Delta G_d^{\ddagger}/RT}$$

where $\Delta G_d^{\ddagger}$ is the chemical free enthalpy of activation of de-electronation. But what is the "concentration" of the metal atoms in the electrode? If we instead use activity, then we might expect to insert the activity of a pure metal, namely one. But if a metal atom is to detach from the surface, it should have a higher energy than a bulk atom: first, there is the surface free energy that it possesses compared to the bulk and second, surfaces are not vicinal or flat. The most likely atom to detach is an atom at a kink site of an edge between two terraces. Let us denote its activity $\bar{a}_{kink}$ in mole m^{-2}. So we should write[2]

$$r_d^0 = \nu_d \, \bar{a}_{kink} \, e^{-\Delta G_d^{\ddagger}/RT}$$

Now if we also account for the electric field in the double layer, we must have

$$r_d = r_d^0 \, e^{+\alpha n F \Delta \phi_{eq}/RT}$$
$$= r_d^0 \, e^{(1-\beta) n F \Delta \phi_{eq}/RT} \tag{14.3.1}$$

where α is the fraction of the total distance travelled by the metal ion from the metal surface until it reaches the point in the double layer where there is a maximum in the free enthalpy. Rather obviously $\alpha = 1 - \beta$. Much of the corrosion literature calls α the *symmetry*

---

[2]West (see Further Reading in Appendix D) prefers to write the backward reaction of (14.1.1) in the more illustrative form

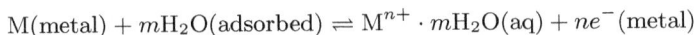

$$M(metal) + mH_2O(adsorbed) \rightleftharpoons M^{n+} \cdot mH_2O(aq) + ne^-(metal)$$

to indicate that a surface metal atom reacts with m water molecules adsorbed on the surface and migrates through the double layer as a hydrated ion. Water has the wonderful ability to point its charged ends towards ions effectively screening their electric field and indeed stabilising charged or zwitterionic species that would otherwise be unstable in the vacuum (A. T. Paxton and J. B. Harper, *Mol. Phys.*, **102**, 953 (2004), *ibid.* p. 1981). On the other hand, if the M^{n+} ion is carried across wreathed in a number of anions such as OH$^-$, then the notion that the work done against the field is $ne\Delta\phi$ becomes untenable as the charge on the migrating object is not simply ne. This must have a large role to play in the structure of the activated complex and in deciding what is the free enthalpy of activation of the reaction (14.1.1). A further complication that I shall ignore is that if we take this more realistic picture of the de-electronation, or metal oxidation, reaction, then it is no longer described by first-order kinetics. In that case, the rate coefficient has different dimensions to k_1. In addition, it is not clear what charge to associate with a screened cation so the electric work is also problematic. Such complications are best avoided, especially considering the comment at the end of the previous section.

factor (or even "transmission coefficient" — which is a phrase I have already used for something else at the end of Section 13.7, or "transfer coefficient"). The equilibrium de-electronation current density is

$$i_d^{eq} = r_d\, nF \qquad (14.3.2)$$

The signs in (14.2.1) and (14.3.1) make sense. If the metal is noble, for example, copper, and hence positively charged with respect to the electrolyte, then $\Delta\phi_{eq} > 0$ and so the barrier is raised for the electronation (metal deposition — reduction) and lowered for de-electronation (metal oxidation). This situation is illustrated by the dotted and solid markings in Figure 14.1 (ignore the dashed for now). Conversely, if the metal is base, for example, magnesium, it is negatively charged with respect to the electrolyte, then $\Delta\phi_{eq} < 0$ and so the barrier is lowered for the electronation and raised for de-electronation. This sounds counterintuitive but see Footnote 4 of Section 9.3, and note that this is a discussion of the *equilibrium* rates of electronation and de-electronation — there is neither overall reduction nor oxidation occurring. In the following section, we see how a positive "overpotential" applied to the metal will upset the equilibrium and drive the oxidation of metal atoms to their ions.

14.4   Further Interpretations of the Symmetry Factor: Overpotential

A simple, and often described, interpretation of the symmetry factor, α, arises from a simplified version of Figure 14.1, sketched in Figure 14.2. Imagine that the solid and dashed lines merge at the right. This implies that the electric potential shift due to the electric field is entirely accomplished by the change of potential on the electrode. In fact, with a view to what follows in Chapter 15, we have allowed, by the use of dashed markings in Figure 14.1, an interphase potential difference of an amount $\Delta\phi$ volts: not necessarily the equilibrium potential, $\Delta\phi_{eq}$, since in a general case the electrode may be connected to power supply. Imagine then that the electrolyte is very concentrated so that any change in potential in the solution is screened out by counter ions. Consider the electronation reaction in terms of Figure 14.2. The M^{n+} ion moving from right to left encounters a barrier in equilibrium shown by the maximum in the solid

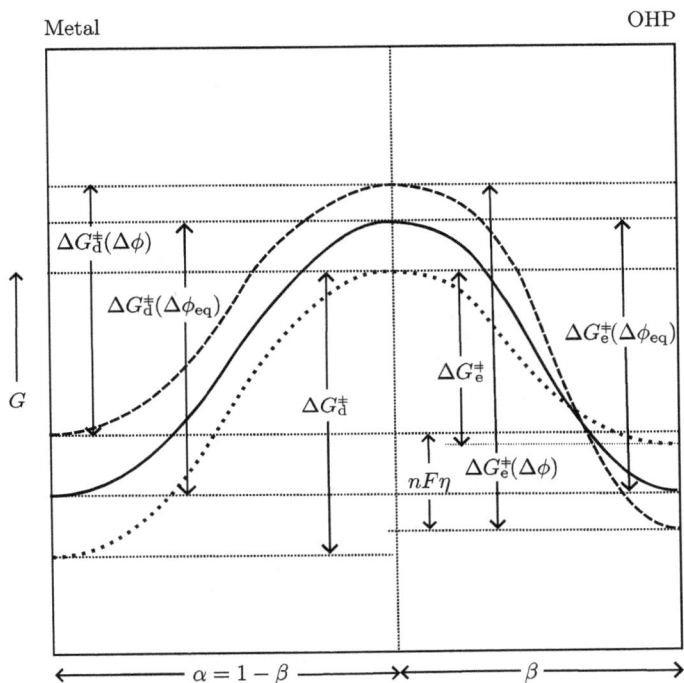

Fig. 14.1. A cartoon of the free enthalpy profile over the reaction coordinate in which an M^{n+} ion is carried across the double layer from right to left, from the OHP to the metal, where it picks up n electrons and is reduced to an M atom; conversely, an M atom leaves n electrons behind at the electrode and is carried to the OHP. The dotted line represents the zero field case, $\Delta\phi = 0$, in which the electric potential is the same in the metal and at the OHP. The solid line corresponds to the equilibrium of (14.1.1), $\Delta\phi = \Delta\phi_{eq}$. If, *in addition* a potential difference $\eta = \Delta\phi - \Delta\phi_{eq}$ is applied across the double layer (say, by attaching the electrode to a battery), then the free enthalpy profile is as shown by the dashed line. In either case, the potential difference alters the free enthalpy difference between the minima at the left and right by an amount $nF\Delta\phi$. In effect, the curves are "tilted" by an amount $nF\Delta\phi$; this alters the height of the maximum — not by the full amount of tilt, $nF\Delta\phi$, but by roughly half that if the maximum is roughly half way along the reaction coordinate. In equilibrium, we have the free energy of product and reactant equal and hence, $\Delta G_d^{\ddagger}(\Delta\phi_{eq}) = \Delta G_e^{\ddagger}(\Delta\phi_{eq})$ as shown. Symmetry factors, α and β, are shown for a total reaction coordinate of one.

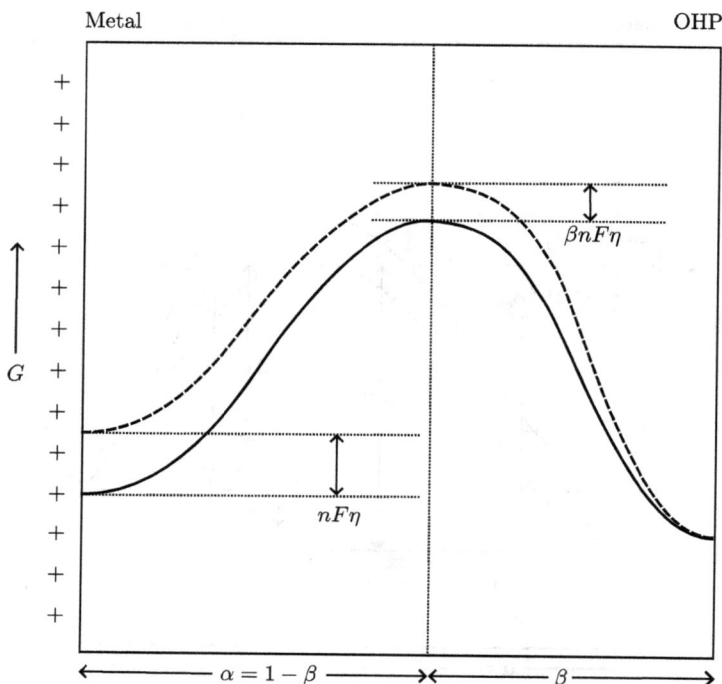

Fig. 14.2. A simpler representation of Figure 14.1 in which the overpotential only "tilts" the free enthalpy profile at the electrode side — it is "hinged" at the right.

curve. By how much is that barrier raised if the free enthalpy of the ion on arrival at the metal is raised by the application of an additional potential difference?

$$\eta = \Delta\phi - \Delta\phi_{eq} \qquad (14.4.1)$$

You can see that the nearer the barrier is to the OHP side of the double layer, the less it is raised; by linear interpolation, the increase in free enthalpy of activation is $\beta n F \eta$ as shown. Compare this with the argument that led to Equation (14.2.1). Conversely, if the M^{n+} ion is moving from left to right, having left its electrons behind in the metal, then the additional electric field supplied by the "over-potential", η, only assists it as far as the maximum, after which it is downhill anyway; hence the rate is enhanced as in (14.3.1) by only that fraction $\alpha = 1 - \beta$ of the total distance travelled times the increase in free enthalpy, $nF\eta$, of the ion at the electrode surface.

A third interpretation is mathematical. In reference to Figure 14.1, we may regard $\Delta G_e^{\ddagger}(\Delta\phi)$ and $\Delta G_d^{\ddagger}(\Delta\phi)$ as functions of the electric field, or equivalently the potential difference $\Delta\phi = \phi_m - \phi_s$ across the double layer (not necessarily the equilibrium potential difference, $\Delta\phi_{eq}$). Then the chemical activation free enthalpies are $\Delta G_d^{\ddagger}(0)$ and $\Delta G_e^{\ddagger}(0)$.[3] We now expand the activation free enthalpies at an arbitrary $\Delta\phi$ about the equilibrium. To first order,

$$\Delta G_d^{\ddagger}(\Delta\phi) = \Delta G_d^{\ddagger}(\Delta\phi_{eq}) - \alpha nF\eta \qquad (14.4.2a)$$

$$\Delta G_e^{\ddagger}(\Delta\phi) = \Delta G_e^{\ddagger}(\Delta\phi_{eq}) + \beta nF\eta \qquad (14.4.2b)$$

in which $\eta = \Delta\phi - \Delta\phi_{eq}$ is called the *overpotential*. For now, α and β are just positive numbers. The signs reflect that an increase in electric potential at the electrode *lowers* the barrier for de-electronation, while it *raises* the barrier for electronation. By the Taylor expansion, the symmetry factor is interpreted as the first-order change in activation free enthalpy due to a change in electrode potential:

$$\alpha = -\left.\frac{\partial\Delta G_d^{\ddagger}}{\partial\Delta\phi}\right|_{\Delta\phi=\Delta\phi_{eq}} \qquad (14.4.3a)$$

$$\beta = \left.\frac{\partial\Delta G_e^{\ddagger}}{\partial\Delta\phi}\right|_{\Delta\phi=\Delta\phi_{eq}} \qquad (14.4.3b)$$

If we were interested in the equilibrium as in Sections 14.2 and 14.3, then we'd expand instead about the zero electric field free enthalpies and in (14.4.2), we would replace the left-hand sides by $\Delta G_d^{\ddagger}(\Delta\phi_{eq})$ and $\Delta G_e^{\ddagger}(\Delta\phi_{eq})$, and on the right-hand sides, we'd have $\Delta G_d^{\ddagger}(0)$ and $\Delta G_e^{\ddagger}(0)$ and replace η with $\Delta\phi_{eq}$:

$$\Delta G_d^{\ddagger}(\Delta\phi_{eq}) = \Delta G_d^{\ddagger}(0) - \alpha nF\Delta\phi_{eq}$$

$$\Delta G_e^{\ddagger}(\Delta\phi_{eq}) = \Delta G_e^{\ddagger}(0) + \beta nF\Delta\phi_{eq}$$

that is, we now consider the dotted and solid not the solid and dashed lines in Figure 14.1. Then you'd see that we obtain the rates (14.2.1)

---

[3]For simplicity, we suppressed the (0) in Sections 14.2 and 14.3.

and (14.3.1) confirming that the first derivatives are indeed the symmetry factors as we have defined them, aside from the subtlety of exactly where we evaluate the derivatives in (14.4.3) which doesn't matter to lowest order. Finally, I can point out that, in reference to Figure 14.1, we have

$$\Delta G_{\mathrm{d}}^{\ddagger}(\Delta\phi_{\mathrm{eq}}) = \Delta G_{\mathrm{e}}^{\ddagger}(\Delta\phi_{\mathrm{eq}})$$

which is the equilibrium free enthalpy of activation of the redox reaction (14.1.1). We also see from Figure 14.1 that

$$\Delta G_{\mathrm{d}}^{\ddagger}(\Delta\phi) - \Delta G_{\mathrm{e}}^{\ddagger}(\Delta\phi) = nF\eta$$

By substitution into (14.4.2), there results

$$\alpha + \beta = 1$$

which is the required property of the symmetry factors.

In this section, I have covered two new matters, both of which will take centre stage in what follows, and are particularly prominent in the theory of corrosion. I have extended the equilibrium considerations of Sections 14.2 and 14.3, to a more general situation in which the coupon of metal is not merely inserted in the solution of its ions, it is also connected to an electrical source or drain. This introduces the *overpotential*, η, in volts. Second, you have the following three ways of thinking about the symmetry factors.

(i) α and β represent a fraction of the total amount of electrical work, $nF\Delta\phi$, that has to be done on the ion in de-electronation and electronation, respectively, reflecting the idea that once the ion reaches that point in its travel at which it encounters the activation free enthalpy maximum, from that point on the journey is "downhill" anyway so no further electrical work needs to be done.

(ii) There is a simple tilting picture as sketched in Figure 14.2. If a free enthalpy profile, such as, say, the solid curve in Figure 14.1, is held fixed at the solution OHP and the left-hand end is pivoted about that point, then if the left-hand point is shifted by $nF\eta$, the maximum in the curve is shifted by a fraction β of that if it is situated a fraction β of the total width of the the double layer from the outer Helmholtz plane. Similarly, if the curve is pivoted from the left and the solution potential is shifted by $nF\eta$, then the maximum changes by only the fraction α of that.

(iii) The first-order expansion of the activation free enthalpy (14.4.2) clearly defines the symmetry factors α and β as in (14.4.3), namely they are minus the equilibrium first-order change in the chemical activation free enthalpy for de-electronation, $\Delta G_d^{\ddagger}$, with potential and the equilibrium first-order change in the chemical activation free enthalpy for electronation, $\Delta G_e^{\ddagger}$, with potential, respectively. This definition with respect to Taylor expansion about equilibrium ensures the condition $\alpha + \beta = 1$ which otherwise we regard as a geometric construction as in Figure 14.1.

14.5 Equilibrium Exchange Current Density

I write out Equations (14.2.2) and (14.3.2) in full, following the derivations in the previous sections:

$$i_d^{eq} = nF\nu_d \, \bar{a}_{kink} \, e^{-\Delta G_d^{\ddagger}/RT + \alpha nF\Delta\phi_{eq}/RT} \qquad (14.5.1a)$$

$$i_e^{eq} = nF\nu_e \, \bar{c}_{M^{n+}} \, e^{-\Delta G_e^{\ddagger}/RT - (1-\alpha)nF\Delta\phi_{eq}/RT} \qquad (14.5.1b)$$

These are the *equilibrium* electronation and de-electronation current densities, and on account of detailed balancing, they must be numerically identical.[4] This leads us to a central quantity in electrochemistry, namely the *equilibrium exchange current density*:

$$i_0 = i_e^{eq} = i_d^{eq} \qquad (14.5.2)$$

In view of all the uncertainties (and the twenty-odd pages of text) leading to Equations (14.5.1), you may be either happy or depressed

---

[4]I do not use plus or minus signs to indicate the direction of current flow — that is obvious from the context. Often, i_d is called the "oxidation current density", i_{ox}, and i_e the "reduction current density", i_{red}, and if signs *are* used, then it is conventional to take i_{ox} as positive and i_{red} as negative. Remember that if you reduce a species, then you donate electrons to it and that oxidation is the taking away of electrons.

to learn that we never need to write them down again! It is enough that we have "understood" processes at the atomic level that are occurring in equilibrium at the surface of a metal immersed in a solution of its cations. An important consequence of the appearance of the activity of a metal atom at a kink site in (14.5.1a) is the role that surface *microstructure* plays. A perfect, ideal low-index surface will possess a very different equilibrium current density compared to a real metal surface. This is particularly important in corrosion in assessing the roles of emerging dislocations and grain boundaries and the different crystal faces that a typical polycrystalline metal surface presents to the outside.[5] All the same, from now on we accept the reality of an exchange current density and I will show in what follows that this is a *measurable* property of a single electrode and moreover that even the crazy *symmetry factor* can be shown to be a measurable reality. Exchange current density may vary from about 10,000 A m^{-2} to as little as 10^{-5} A m^{-2}. It is surprising that there is a nine orders of magnitude variation across different metals.

14.6   Interpretation of the Equilibrium Potential

Before leaving Equations (14.5.1), let us set the two right-hand sides equal to each other, as indeed they are in equilibrium. Then we get

$$nF\Delta\phi_{eq} = \Delta G - G^0 \qquad (14.6.1)$$

in which we may regard as "constant"

$$G^0 = -RT \ln \frac{\nu_e \, \bar{c}_{M^{n+}}}{\nu_d \, \bar{a}_{kink}}$$

and, in reference to Figure 14.1,

$$\Delta G = \Delta G_d^{\ddagger} - \Delta G_e^{\ddagger}$$
$$= G_M - G_{M^{n+}}$$

---

[5]In their Chapter 15, Schmickler and Santos (see Further Reading in Appendix D) have some much deeper insights into the atomic scale processes of metal deposition and wastage and show some detailed models of how these impact on the kinetics.

is a sort of free enthalpy change for the reaction (14.1.1). Here are some points to note:

(i) It is the equality of (14.5.1a) and (14.5.1b) that establishes the value of $\Delta\phi_{eq}$, since all other quantities are fixed by the nature of the experiment. This tells us something fundamental about the equilibrium electric potential that is established across the double layer when a coupon of pure metal is placed into a solution of its ions.

(ii) The free enthalpy difference, ΔG, is not the thermodynamic free enthalpy change of the reaction (14.1.1) because it involves the free enthalpy change of the specific reaction in which a hydrated metal cation is transferred through an electric field to become adsorbed at a surface kink site of the metal.

(iii) From that perspective, we may attempt a thought experiment. We take a metal atom to the Bockris point outside a metal vacuum interface; this costs us free enthalpy per atom (or metal cohesive energy) $-\Delta G_{coh}^{M}$; we strip off n electrons, which costs us the first n ionisation energies; we return these electrons to the metal, gaining us n times the metal's work function; we transfer the M^{n+} cation through the vacuum from the Bockris point of the metal to the Bockris point of the solution; the work done is $-ne\,(\psi_m - \psi_s)$; finally, we plunge the cation into the solution requiring an amount of work $-\Delta_{sol}^{r}G(M^{n+})$ which is minus the real free enthalpy of solvation of the M^{n+} ion (see Chapter 7). The result (per cation) is

$$\Delta G \approx \Delta G_{coh}^{M} - (I_1 + \cdots + I_n) + nW_M$$
$$+ ne\,(\psi_m - \psi_s) + \Delta_{sol}^{r}G(M^{n+}) \qquad (14.6.2)$$

This is an approximate formula because it does not account for the difference in free enthalpy of bulk atoms and cations and those at the metal surface and outer Helmholtz plane, respectively, and it does not account for the electric charges of the two phases.

(iv) All the same, this formula throws some light on the factors advanced by Fawcett (see Section 9.5) that determine the position of a particular metal in the electrochemical series. If we associate $\Delta\phi_{eq}$ with standard half cell electromotance, $\mathcal{E}_M^{\circ}$, and if we then compare (14.6.1) and (14.6.2), we see that what controls a metal's position in the electrochemical series are indeed the free enthalpy of solvation of the M^{n+} cation, the free enthalpy (cohesive energy) of the metal and the ionisation energies of the metal atom as stated by Fawcett. Furthermore, we find that further factors are involved, namely the work function of the metal and difference in Volta potentials between the metal and the electrolyte.

14.7 Nernst Equation from the Point of View of Kinetics

Please return to Section 9.6. Reconsider the half cell reaction

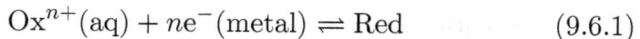

$$Ox^{n+}(aq) + ne^-(metal) \rightleftharpoons Red \tag{9.6.1}$$

Let us suppose we can write its equilibrium exchange current density in terms of electronation and de-electronation rate coefficients in first-order kinetics, the rate of the forward and back reactions being given by the rate coefficient times the *activity* of the species. The electronation and de-electronation (reduction and oxidation) rate coefficients are, by reference to (14.2.1) and (14.3.1),

$$k_e = \nu_e\, e^{-\Delta G_e^{\ddagger}/RT}$$

$$k_d = \nu_d\, e^{-\Delta G_d^{\ddagger}/RT}$$

and the equilibrium exchange current density is

$$i_0 = i_e^{eq} = i_d^{eq} = nFk_e a_{Ox}\, e^{-(1-\alpha)nF\Delta\phi_{eq}/RT}$$

$$= nFk_d a_{Red}\, e^{\alpha nF\Delta\phi_{eq}/RT} \tag{14.7.1}$$

Rather as in Section 14.6, we find that the symmetry factor cancels when we resolve this equality into

$$\Delta\phi_{eq} = \frac{RT}{nF}\ln\frac{k_d}{k_e} + \frac{RT}{nF}\ln\frac{a_{Ox}}{a_{Red}}$$

In the standard state in which the activities are one, the second term on the right-hand side vanishes and we are led to define a standard equilibrium potential:

$$\Delta\phi_{eq}^{\circ} = \frac{RT}{nF}\ln\frac{k_d}{k_e}$$

so bearing in mind that the equilibrium potential difference (14.1.2) is defined for a particular concentration of M^{n+} ions in solution, we can express the equilibrium electrode potential difference as

$$\Delta\phi_{eq} = \Delta\phi_{eq}^{\circ} + \frac{RT}{nF}\ln\frac{a_{Ox}}{a_{Red}} \tag{14.7.2}$$

This is the half cell or *single electrode* form of the Nernst equation which we already encountered in Section 9.2. You can recover Equation (9.6.2) which is the usual form of the Nernst equation expressed with respect to the standard hydrogen electrode (SHE) simply by imagining that our half cell is connected to an SHE as in Figure 9.1. In that case, add to both sides of (14.7.2)

$$(\phi_{Cu'} - \phi_m) + (\phi_s - \phi_{Pt}) + (\phi_{Pt} - \phi_{Cu})$$

if we are using copper leads, and you end up with

$$\mathcal{E} = \mathcal{E}^{\circ} + \frac{RT}{nF}\ln\frac{a_{Ox}}{a_{Red}} \tag{9.6.2}$$

which is the Nernst equation expressing the equilibrium single electrode potential on the SHE scale as a function of the activities.

What we've achieved in this section is to derive the Nernst equation from the kinetics of the redox reaction, rather that the thermodynamics, by setting the equilibrium condition that equates the forward and backward current densities. I repeat that the symmetry factor has cancelled, which indicates that there is *no equilibrium property* or measurement that can be used to extract symmetry factors — they are a strictly *kinetic* quantity (which we will see can be measured). I hope that this perspective on the Nernst equation provides additional insight into the thermodynamic derivation in Chapter 9. It also serves to point out that there exists a single electrode form (14.7.2) of the Nernst equation, albeit an expression of an unmeasurable quantity, $\Delta\phi_{eq}$.

14.8 Measurement of the Exchange Current Density

In Figure 2.4, Chapter 2, I sketched the "next simplest electrochemical cell". In Figure 14.3, I sketch the "next-next simplest electrochemical cell". On the left is the familiar standard hydrogen electrode (see Section 9.4). This is connected through a high impedance voltmeter

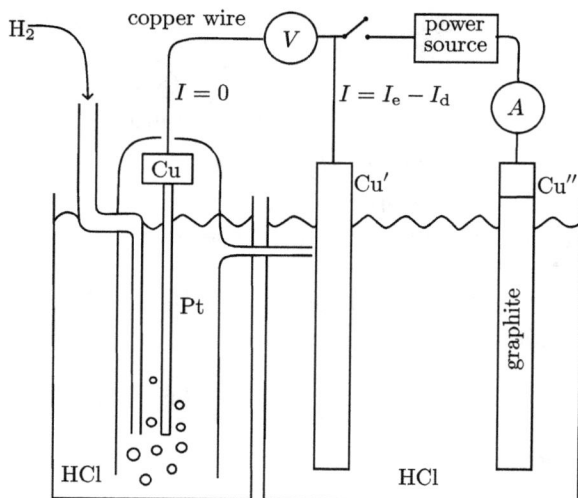

Fig. 14.3. A standard hydrogen electrode connected through a voltmeter to a copper electrode immersed in hydrochloric acid electrolyte. The "counter electrode" is there to allow the current to be controlled.

(to prevent any current flowing) to a copper test electrode which is immersed in hydrochloric acid (HCl). The finger-like probe is a "Luggin capillary"; its purpose is to minimise any electrical resistance in the electrolyte so that the SHE measures exactly the potential of the electrolyte very close to the interphase, relative to the SHE. With the switch open, this is equivalent to the electrochemical cell in Figure 9.1, and you do not for a moment imagine that this gives us a direct measure of $\Delta\phi_{eq}$ because of the usual bloody mindedness of electrochemical experiments causing additional potential differences to intervene. With no current flowing, the potential difference measured at the voltmeter is

$$\mathcal{E}_{eq} = (\phi_{Cu'} - \phi_{HCl}) + (\phi_{HCl} - \phi_{Pt}) + (\phi_{Pt} - \phi_{Cu})$$
$$= \Delta\phi_{eq} + \Delta\phi_{SHE} + (\phi_{Pt} - \phi_{Cu})$$

When the switch is closed, the cell has *three electrodes* in order to be able to control the current $I = I_e - I_d$ by varying the voltage of the power supply independently of the voltage on the SHE. The right-hand side electrode is a so-called auxiliary or *counter electrode*. It is made of an inert conductor such as graphite so that the only

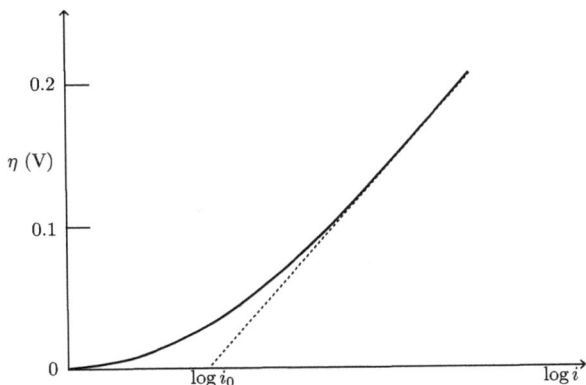

Fig. 14.4. Overpotential as a function of log current density in the cell of Figure 14.3. By extrapolation, the exchange current density is measured.

electrochemical reaction involves the electrolyte, for example, the oxidation of chloride ions to chlorine gas. If the power source is turned off, then the voltmeter records the equilibrium potential difference, $\mathcal{E}_{eq}$. As the voltage in the power source is ramped up, the current, I, is measured at the ammeter and the voltage recorded at the voltmeter is

$$\mathcal{E} = \Delta\phi + \Delta\phi_{SHE} + (\phi_{Pt} - \phi_{Cu}) - IR$$

where $\Delta\phi$ is the non-equilibrium potential difference across the double layer of the test electrode and R is the resistance of the electrolyte between the tip of the Luggin capillary and the test, or *working* electrode. The point is that we can eliminate the "IR-drop" by means of careful assembly and glass blowing. We may also take it that as the voltage in the power source and the resulting current are ramped up, the standard potential of the SHE and the inner potential difference at the copper–platinum interface in the SHE remain constant.

If the logarithm of the current density, i (I divided by the surface area of the test electrode), is plotted against the voltage, $\mathcal{E} - \mathcal{E}_{eq} = \Delta\phi - \Delta\phi_{eq} = \eta$, a plot as Figure 14.4 is typically obtained. An extrapolation to zero voltage permits a measurement of the equilibrium exchange current density, i_0. This is an elementary example of *voltammetry* which we encountered in Section 11.3.

So the point is that i_0 is *measurable*. On the other hand, it is not possible to manipulate Equations (14.5.1) as we could to find a simple form for $\Delta\phi_{eq}$. Indeed, unlike $\Delta\phi_{eq}$, i_0 is seen to depend on

Table 14.1.   Measured data for a few redox couples. Exchange current density, i_0, $\mathcal{E}_M^\circ$ from Table 9.3 and Tafel slope.

Redox system	Metal	i_0 (Am^{-2})	$\mathcal{E}_M^\circ$ (V/SHE)	Tafel slope, b (V/decade)
Ag$^+$|Ag	Ag	10^4	+0.80	0.03–0.12
Cu^{2+}|Cu	Cu	1–10	+0.34	≈0.06
Zn^{2+}|Zn	Zn	10^{-3}–10^{-1}	−0.76	0.03–0.06
Fe^{2+}|Fe	Fe	10^{-5}–10^{-4}	−0.44	0.05–0.08
H$^+$|H$_2$	Pt	10^2	0	≈0.12
H$^+$|H$_2$	Fe, Cu	10^{-3}–10^{-2}	0	≈0.12
H$^+$|H$_2$	Zn	10^{-7}	0	≈0.12
H$^+$|H$_2$	Pb	10^{-9}	0	≈0.12
O$_2$|OH$^-$	Pt	10^{-6}	+0.4	0.10–0.15
O$_2$|OH$^-$	Fe	≈ 10^{-10}	+0.4	>0.12

the height of the activation barrier and on at least one of the vibrational and concentration terms. It must be because of the exponential dependence on the activation free enthalpy that there is a nine order of magnitude variation in i_0 across a range of metals. Mind you, that still translates into a factor of about twenty in comparing the smallest and largest activation barrier, which is rather puzzling. The frequency and concentration prefactors cannot vary that much between metals. This implies that the reality is more complicated than the picture I have drawn. Table 14.1 shows some exchange current densities in water at 25°C. (Please just look at the first four lines and four columns of data for now.) It is also puzzling that there is a variation of ten or a hundred even for the same metal and electrolyte. This may reflect the importance of the state of the surface as I have mentioned already. There is a rough correlation between exchange current density and equilibrium potential (Table 9.3). The more noble metals have the largest exchange current density.

14.9   Interpretation of the Potential of Zero Charge

I have said that an isolated piece of metal placed in water, or in a solution of its ions, will spontaneously acquire a charge. As a consequence, the solution acquires an equal and opposite charge and a potential difference, $\Delta\phi_{eq}$, is developed across the interphase between metal and solution (electrolyte). This is of course accompanied by an electric field across the interphase, which we have taken as constant. I have then asserted that the conditions of charge and the potential difference can be modified by connecting the metal to ground, or to a source or drain of charge such as a battery. By connecting the metal to a negative or positive terminal of a battery, we effectively supply an *overpotential*, η (itself a potential difference),

$$\eta = \Delta\phi - \Delta\phi_{eq} \tag{14.4.1}$$

and $\Delta\phi$ is the resulting potential difference across the interphase. Therefore, in an experiment such as described in Section 11.2, I can find that overpotential which results in zero potential difference across the interphase, and in that condition the metal is neutral. Since the metal now carries zero charge, I have arrived at that overpotential which corresponds to the potential of zero charge that was encountered first in Section 11.2. As we vary $\mathcal{E}_a$, at $\sigma = 0$, we have a maximum in the surface excess free enthalpy, but we also have identified that applied e.m.f. which results in $\Delta\phi = 0$. In that case, we have, from (14.4.1) with $\Delta\phi = 0$,

$$\eta_{pzc} = -\Delta\phi_{eq}$$

However, the PZC as defined by (11.2.4) refers to minus the electrode surface charge so we need to reverse the sign of the applied e.m.f. and we expect to have

$$\phi_{pzc} = \Delta\phi_{eq}$$

You have seen in Section 14.7 that the equilibrium potential difference, $\Delta\phi_{eq}$, is related to the standard electrode potential if both are measured with respect to the standard hydrogen electrode and if the metal ions in the electrolyte are at unit activity. This implies a relation between the PZC and the equilibrium standard electrode potential. So I can combine data from the electrochemical series, Table 9.3, with some potentials of zero charge from Table 11.1. You do not expect equality between these two because although both are referred to the SHE, the electrolytes in the measurements of the PZC are not at unit activity of the metal ions. The comparison is

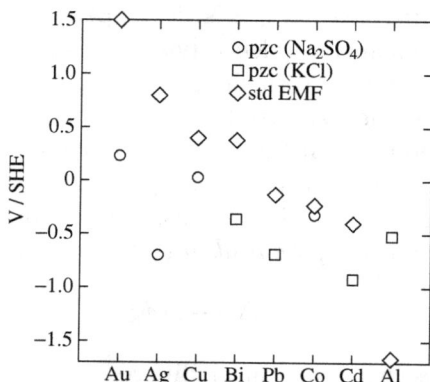

Fig. 14.5. A plot comparing the standard e.m.f. to the potential of zero charge for a number of metals.

presented graphically in Figure 14.5. The correlation is not wholly convincing, but the trend is there.

All other things being equal, the PZC should be proportional to the metal's work function as has been shown by Trasatti.

14.10   Final Remarks on the Baseness and Nobility of Metals

I have now attempted to attach some physical substance to the electrochemical series — perhaps the most important feature of electrochemistry.

(i) I have remarked at the end of Section 9.5 upon the essential fact that *some metals are base* and there is nothing the materials designer can do about this depressing thought. In Chapter 17, we come to the subject of passivation and the existence of "stainless" metals.

(ii) Fawcett has given a list of thermodynamic properties that influence the position of a metal in the electrochemical series (Table 9.3), which I repeat and extend at point (iv) in Section 14.6.

(iii) I have now indicated a possible correlation between potential of zero charge, equilibrium potential difference and position in the electromotive series. Application of the Nernst equation should bring these into closer coincidence if the conditions of the experiment are known.

(iv) I have also pointed out at the end of Section 14.8 that there is a rough correlation between position in the electromotive series and exchange current density, although the latter is also affected by the catalytic properties of the electrode metal.

Chapter 15

Overpotential and the Butler–Volmer Equation

15.1 The Butler–Volmer Equation

We are ready to derive the Butler–Volmer equation. By now you are aware that an isolated metal in a solution of its ions inevitably sets up an equilibrium potential difference between metal and solution which we have called $\Delta\phi_{eq}$ (see Equation 14.1.2). We have also seen that it represents an *unmeasurable* surrogate for the standard electrode potential, $\mathcal{E}_M^\circ$, on the SHE scale (see the electromotive force series, Table 9.3) once adjusted to standard conditions using the Nernst equation. In addition, I have indicated that the equilibrium will be upset if an additional overpotential, η, is applied to the electrode (or the electrolyte). To my mind, there is an analogy between free charge and bound charge in classical electrostatics, and overpotential and equilibrium potential in electrochemistry. In the study of ponderable electric media, we learn that an applied electric field will cause matter to polarise and that so-called bound charge will appear within the dielectric and on its surfaces due to atomic scale charges becoming shifted from equilibrium as a result of the applied field. The applied field of course is created by us by the manipulation of electronic circuitry, for example, by moving charge onto the plates of a capacitor. In that sense, we call the charge that we manipulate and control the *free charge*, and the *bound charge* or *inaccessible* charge is not measurable and is out of our control being the inevitable result of matter having an electric susceptibility. In the same way, we control the *overpotential* in an electrochemical experiment, but we do not control the unmeasurable potential drops across the interphase which result from atomic scale movements of charge, which again are dictated by the properties of the electrode and the electrolyte which are outside our control. Indeed, just as in electrostatics, it is not clear to me that we may always unambiguously identify which part of $\Delta\phi$ is equilibrium potential and which part is overpotential.

However, let us take the point of view that the electrode potential, $\Delta\phi$, *can* be divided into the equilibrium part and the overpotential, as introduced in (14.4.1):

$$\Delta\phi = \Delta\phi_{eq} + \eta$$

The electrode doesn't care about this division — to it $\Delta\phi$ is just a potential drop across the interphase, but to us it's important because η is the part that we can control: the rest comes along for the ride.

If the potential drop is $\Delta\phi_{eq}$, then the interphase is in equilibrium; there is no net current flowing and $i_d^{eq} = i_e^{eq} = i_0$. A current *will* flow if an overpotential is applied, and the question is what is then the resulting current density, $i = i_d - i_e$. It is the convention that positive charge flowing from electrode to electrolyte is designated a positive current, see Footnote 4 of Section 14.5. We can calculate $i = i_d - i_e$ using an obvious extension of Equation (14.7.1) to non-equilibrium by replacing $\Delta\phi_{eq}$ with the non-equilibrium potential difference across the interphase, $\Delta\phi$:

$$i = nFk_d a_{Red}\, e^{\alpha nF\Delta\phi/RT} - nFk_e a_{Ox}\, e^{-(1-\alpha)nF\Delta\phi/RT}$$

$$= nFk_d a_{Red}\, e^{\alpha nF\Delta\phi_{eq}/RT} e^{\alpha nF\eta/RT}$$

$$- nFk_e a_{Ox}\, e^{-(1-\alpha)nF\Delta\phi_{eq}/RT} e^{-(1-\alpha)nF\eta/RT}$$

$$= i_d^{eq} e^{\alpha nF\eta/RT} - i_e^{eq} e^{-(1-\alpha)nF\eta/RT}$$

$$= i_0 \left(e^{\alpha nF\eta/RT} - e^{-(1-\alpha)nF\eta/RT} \right)$$

The wonderful thing is that it doesn't really matter how I formulate the equilibrium electronation and de-electronation current densities — for example, I could have used (14.5.1) — or how I argue what exactly is the right way to express the "frequency" and "concentration" terms because by pulling out the equilibrium potential

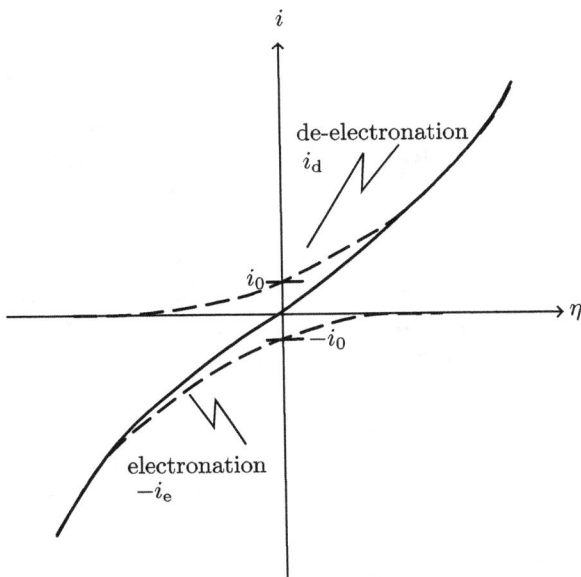

Fig. 15.1. Current density versus overpotential in accord with the Butler–Volmer equation. The two terms of Equation (15.1.1) are shown as broken lines and their sum as the solid line.

from $\Delta\phi$ and using $i_{\mathrm{d}}^{\mathrm{eq}} = i_{\mathrm{e}}^{\mathrm{eq}} = i_0$ *in equilibrium*, all the mess cancels and I can fulfill my promise at the end of Section 15.3 that we will never need them again.

Just because it's so important I'll repeat the Butler–Volmer equation:

$$i = i_0\left(e^{\alpha nF\eta/RT} - e^{-(1-\alpha)nF\eta/RT}\right) \tag{15.1.1}$$

in fact, this is what is plotted in Figure 14.4, and in more detail in Figure 15.1. This and the Nernst equation may be the most important statements in electrochemistry, one expressing the kinetics and the other the equilibrium of the interphase. The Butler–Volmer equation is of particular importance in corrosion[1] and in the following

---

[1]It is fascinating that in the new science and engineering of fuel cells, the design is often driven by competition between kinetic and thermodynamic effects which tend to pull in opposite directions. This book would double in length if I went into fuel cells and batteries, but I hope the reader now has enough background to engage in these fields of vital importance to the climate emergency.

few paragraphs, I make some analysis of the limits of small and large overpotential. Points to note in Figure 15.1 are the individual de-electronation and electronation current densities which add to give the total current density, the intersections of i_d and i_e with the ordinate at $\eta = 0$ to indicate their equality to the equilibrium exchange current density, i_0, and the observation that at large magnitudes of the overpotential the total current density merges with i_d if $\eta > 0$ and i_e if $\eta < 0$, which we return to at Equations (15.3.1).

But I should end this section on a strict caveat. We have swept away the details, approximations and imprecise arguments into the exchange current density, i_0, and obtained a clear physical law that can be readily interpreted and which has great power in the study of corrosion, electroplating, galvanising and so on. But as Schmickler and Santos are careful to point out, the validity of the Butler–Volmer equation must be established on a case-by-case basis, and in many situations, it is far from being justified. In some cases, a detailed theory can be cast into Butler–Volmer form; however this may result in apparent and indeed measured symmetry factors that are not between zero and one, for example. So caution is needed.

15.2   The Limit of Small Overpotential

If I expand the exponentials in (15.1.1) to first order, the terms in the symmetry factor cancel and I get simply

$$i = i_0 \frac{nF\eta}{RT}; \quad nF\eta \ll RT$$

You can see by a quick calculation that at 300 K and $n = 1$, this will be good to about a 1% error if the overpotential is less than 0.01 V, which really is quite small. The symmetry factor cannot be measured at such low overpotential.

In this linear regime, the current–voltage relation is *ohmic*. We could write

$$\left(\frac{\partial \eta}{\partial i}\right)_{\text{activity},T} = \frac{RT}{i_0 nF} = \rho\, d \tag{15.2.1}$$

in which ρ is the differential resistivity of the interphase of width d. This throws some light on polarisable and non-polarisable (reversible) electrodes which I first introduced in Chapter 2. This is exactly the resistivity that dictates the resistance in the model of Figure 2.6 of the interphase as a resistor and capacitor in series. An ideally reversible electrode such as the SHE is intended to have $\rho \to 0$ so that it is perfectly "leaky" of current. So it cannot sustain a change in potential difference ensuring that any change in voltage in an experimental cell such as in Figure 2.4 is entirely due to the polarisation of the test electrode and not the reference electrode. For this measurement to be valid, it is also necessary that the test electrode is ideally polarisable, meaning that $\rho \to \infty$. In this way in the model of Figure 2.6, the polarisable interphase is modelled by a capacitor only and the reversible interphase by a resistor only. Recall that in Sections 11.3 and 12.1 we get a brief reference to the modelling of an interphase as an equivalent circuit of resistors, capacitors and indeed inductors.

The *reversibility* of the SHE is clearly reflected in the very high equilibrium exchange current density for proton reduction on platinum which is evident in Table 14.1, Section 14.8. Conversely as is evident from (15.2.1), a signature of a *polarisable* electrode is one having a very small i_0.

15.3 The Limit of Large Overpotential

It is instructive to make a sketch of the current density against overpotential. Note that if $\alpha = 0.5$, then (15.1.1) is

$$ i = i_0 \sinh \frac{1}{2} \frac{nF\eta}{RT} $$

So mathematically, α dictates the asymmetry of the i–η relation under a change in polarity: a reversal in the sign of the voltage. Conversely, if α is not equal to one half, then as shown in Figure 15.2, the i–η relation is not symmetric: the interphase is non-ohmic — it is *rectifying* (like a diode). This admits a direct measurement of the symmetry factor, simply by reversing the polarity in an experiment, as shown in Figure 14.3 and discussed in Section 14.8. This means that at high overpotential, the equivalent circuit modelling of the interphase must include some nonlinear elements.

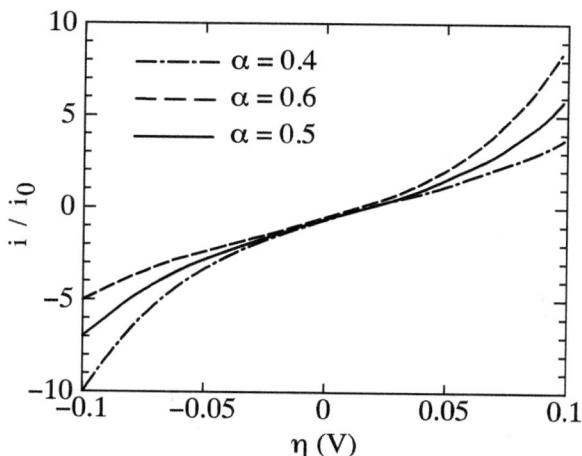

Fig. 15.2. Current density versus overpotential at different values of the symmetry factor.
Source: Adapted with permission from Schmickler and Santos (see Further Reading in Appendix D).

In the limit of large $|\eta|$, either the first or the second term in the Butler–Volmer equation is neglected (see Figure 15.1). In that case the electronation and de-electronation current densities display the asymmetry since

$$i_d = i_0 e^{\alpha n F \eta / RT}; \qquad n F \eta \gg RT \qquad (15.3.1a)$$

$$i_e = i_0 e^{-(1-\alpha) n F \eta / RT}; \qquad -n F \eta \gg RT \qquad (15.3.1b)$$

are only equal if $\alpha = \frac{1}{2}$.

The approximations (15.3.1) are generally correct for $n = 1$ with less than 0.1% error for overpotentials that are greater than 1.2 V in magnitude. Equations (15.3.1) are frequently expressed in the form of the empirical Tafel's law

$$\eta = a \pm b \log \frac{|i|}{i_1} \qquad (15.3.2)$$

using logarithms to the base of 10 and in which i_1 is a constant of unit current density in order to render the argument to the logarithm dimensionless. a and b are the "Tafel coefficients" and b, always positive, is frequently expressed in units of volts per tenfold change in i,

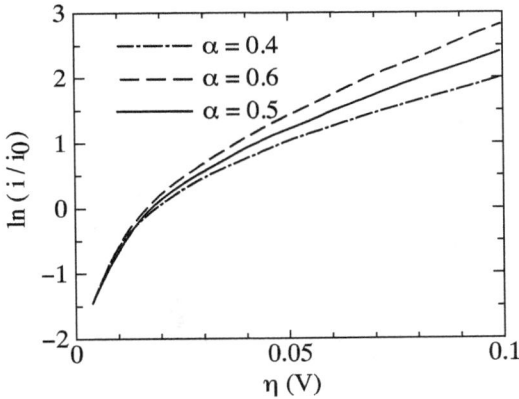

Fig. 15.3.   Current density versus overpotential in a log linear plot at different values of the symmetry factor.

Source: Adapted with permission from Schmickler and Santos (see Further Reading in Appendix D).

or "volts per decade". At de-electronation, we take the plus sign and the anodic overpotential is positive; conversely, the cathodic overpotential is negative and we take the minus sign. Tafel's law predates by some decades the foregoing theory, which may have been developed, as Professor Bilby used to say, "with an eye to the final result". The Tafel b-coefficient is often called the Tafel slope. Some Tafel slopes are given in Table 14.1, Section 14.8. We come to the second two sets of data in that table in a moment.

Figures 15.2 and 15.3 show quantitative graphs of $i-\eta$ and $\ln i-\eta$ relations. Note that for overpotentials greater than about 0.1 V, the graph of $\ln i$ versus η is essentially *linear*. This is taken to be valid almost universally in the treatment of corrosion and leads to the widespread use of the Tafel plot and Evans diagrams, which we come to in the next chapter.

Chapter 16

The Evans Diagram and the Corrosion Potential

16.1 Corrosion

I wish now to turn to corrosion for the remainder of this text. We need to revisit John West's question of why metals corrode at all in the case that the metal is not directly connected to a source or drain of electrons. As I mentioned before, what will happen is that patches of the metal surface will become anodes and patches will become cathodes and current will flow between the two to maintain charge neutrality while the anode wastes away and the cathode supports some reduction reaction — often the reduction of dissolved oxygen, or in acid solutions the reduction of protons to hydrogen gas. This situation is illustrated in Figure 16.1.

You could carry out the following experiment if you have access to two chemicals. Take a big lump of iron (say from a scrapyard) and clean and degrease a flat surface and rough it up a bit with emery cloth. Then make up a solution of water with a little salt (optional) and some potassium hexacyanoferrate and some phenolphthalein indicator. Shake it up a bit to saturate it with dissolved oxygen. You place a big drop of this onto the iron surface, maybe about half an inch or so across. The iron will begin to corrode having chosen a number of patches to become anodes; you will see these patches turning blue as the potassium hexacyanoferrate indicates the appearance of Fe^{++} ions in the water. To conserve charge, cathodic patches also appear and the cathodic reaction is the reduction of oxygen

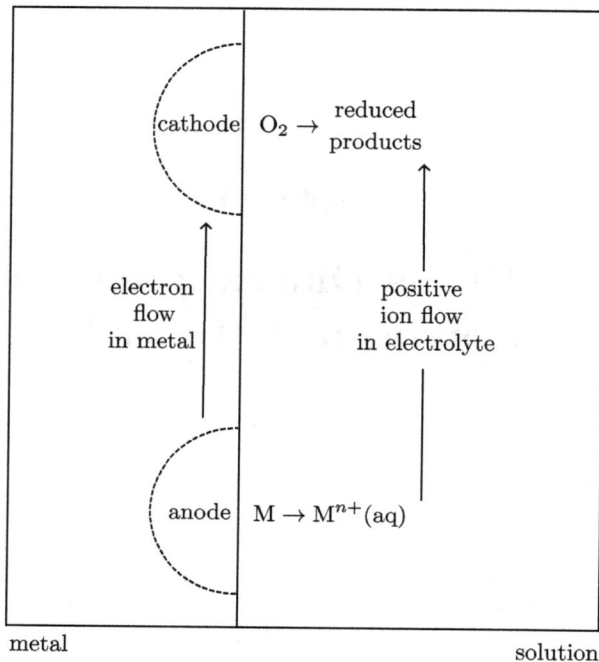

Fig. 16.1.   Corrosion: the metal on the left divides itself into regions of anode and cathode so as to create two half cells and complete an electric circuit as shown. The electrolyte is on the right in which de-electronation takes place at the anode and electronation happens at the cathode (see Figure 1.2 in Chapter 1).

(see (16.2.1)). These patches will show up pink which is an indicator of alkali — the presence of OH^- ions. After a while, the dissolved oxygen will run out and the appearance will change: the blue patches will turn into a ring of blue around the edge of the drop where further oxygen can be supplied from the air in contact with both the water edge and the metal, and the centre of the drop will become a red cathode. Eventually, a ring of rust will appear that separates the outer blue ring and the central pink patch and the experiment will be over.

16.2   Redox Equilibrium

The Nernst equation as we have encountered it up to now applies to electrochemical reactions that actively involve the electrode as it is its atoms that are involved in the so-called *redox reaction* most generally expressed as before, Sections 9.6 and 14.7:

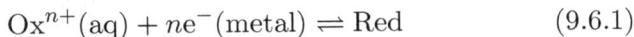

$$Ox^{n+}(aq) + ne^-(metal) \rightleftharpoons Red \qquad (9.6.1)$$

However, it is more general than this in the sense that the species being oxidised or reduced need not belong to the metal electrode. For example, Fe^{++} and Fe^{+++} ions may be in equilibrium in a solution, but if an inert electrode such as platinum is inserted into the solution, then the amounts of the two ions in equilibrium can be affected by the electric potential imposed by a battery on the platinum. This is because in the absence of an electrode, the relative concentrations of Fe^{++} and Fe^{+++} must be fixed by charge neutrality, but the intervention of a metal that might supply or drain electrons can upset this balance. From that point of view, a bulk redox reaction becomes a surface redox reaction, and this is a most important point to bear in mind in what follows.

Let me give an example of each of these. Firstly, return to the electrochemical cell in Figure 9.1. There exists an equilibrium

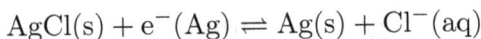

$$AgCl(s) + e^-(Ag) \rightleftharpoons Ag(s) + Cl^-(aq)$$

in which "s" stands for "in solid phase", "Ag" stands for "in the silver electrode" and "aq" is "in aqueous solution", and in this case, the species being reduced (in the forward reaction) and oxidised (in the backward reaction) is an atom of the electrode metal itself. The standard electromotance of this equilibrium can be worked out as

$$\mathcal{E}^\circ = \frac{1}{F}\left(\mu^{\bullet}_{AgCl} - \mu^{\square}_{Cl^-}\right) = \frac{1}{96.48}(-109.8 - 131.1) = -2.5 \text{ V/SHE}$$

where I have used the standard chemical potentials in kJ mol^{-1} at 25°C relative to the hydrogen ion from thermodynamic tables. Application of the Nernst equation allows me to calculate the equilibrium potential at different temperature and concentration of chloride ions in the solution

$$\mathcal{E} = \mathcal{E}^\circ + \frac{RT}{F}\ln\frac{a_{AgCl}}{h_{Cl^-}}$$

Secondly, consider this redox equilibrium

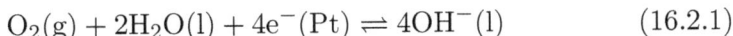

$$O_2(g) + 2H_2O(l) + 4e^-(Pt) \rightleftharpoons 4OH^-(l) \qquad (16.2.1)$$

Here, "g" means "gas", "l" means "liquid" and "Pt" is an inert platinum electrode. So here the electrode plays no part in the reaction *except to act as a source or sink of electrons* and also possibly to catalyse the reaction (increase its rate without itself being altered). Again, I can work out the equilibrium potential of this extremely important reduction of oxygen:

$$\mathcal{E}^\circ = \frac{1}{4F}\left(2\mu_{H_2O}^{\bullet} - 4\mu_{OH^-}^{\square}\right)$$

$$= \frac{1}{385.9}\left(-474.4 + 628.8\right)$$

$$= +0.40 \text{ V/SHE} \tag{16.2.2}$$

Application of the Nernst equation gives us

$$\mathcal{E} = \mathcal{E}^\circ + \frac{RT}{4F}\ln\frac{p_{O_2}}{h_{OH^-}^4}$$

$$= 0.40 + 0.015\log p_{O_2} - 0.06\log h_{OH^-}$$

which includes a change from natural to base-10 logarithms. The partial pressure of oxygen must be in bar. Now you need to know what is meant by *pH*. This is a measure of *acidity* which may be characterised by the presence of H^+ ions (protons) in solution. The measure used is minus the logarithm to the base ten of the concentration (Henrian activity) of hydrogen ions.[1] In terms of the *pH* then,

---

[1]Consider the equilibrium

$$2H_2O \rightleftharpoons H_3O^+ + OH^-$$

The proton does not really exist in water (although there is evidence that it does exist in the double layer, see https://tonypaxton.org/Water.html), by H^+ we really mean the hydronium cation as on the right. This equilibrium expresses the fact that water is always slightly dissociated into ions (that's why it conducts electricity even when pure). Pure water at $25°C$ is *neutral* and the concentration of each ion is equal and such that its activity is 10^{-7} (from observation). We write the equilibrium constant for the dissociation as

$$K_W = h_{OH^-} h_{H_3O^+}$$

assuming the activity of water is one since there are so few ions. We *define*

$$pH = -\log h_{H_3O^+} \quad \text{and} \quad pOH = -\log h_{OH^-}$$

the equilibrium electromotance of the oxygen reduction at a metal surface at 25°C is

$$\mathcal{E} = 1.23 + 0.015 \log p_{O_2} - 0.06 \, pH \qquad (16.2.3)$$

in volts on the standard hydrogen electrode scale. In neutral water with dissolved oxygen in equilibrium with air whose partial pressure is 0.2 bar (20% of one atmosphere), the equilibrium potential for oxygen reduction on metal is

$$\mathcal{E} = 1.23 + 0.015 \log 0.2 - 0.06 \times 7 = +0.8 \, \text{V/SHE} \qquad (16.2.4)$$

In acid solutions, the cathodic reaction is likely to be reduction of hydrogen ions rather than dissolved oxygen. In that case, the reaction is[2]

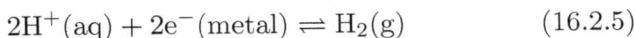

$$2H^+(aq) + 2e^-(metal) \rightleftharpoons H_2(g) \qquad (16.2.5)$$

This is the "hydrogen evolution reaction" (HER) that occurs at the standard hydrogen electrode (SHE) and since the electromotance series is constructed with reference to the SHE, the equilibrium standard electromotance of reaction (16.2.5) is, by construction, zero volts. If the activity of hydrogen ions is not one, then the electromotance is

$$\mathcal{E} = \frac{RT}{2F} \ln \frac{h_{H^+}^2}{p_{H_2}}$$
$$= -2.303 \frac{RT}{F} \left(pH + \frac{1}{2} \log p_{H_2} \right) \qquad (16.2.6)$$

again, the partial pressure of hydrogen gas is in bar, and here and in (16.2.4) we assume the gas is ideal. The number $\ln x / \log x = 2.303$ turns up a lot in electrochemistry calculations. Note, also, that $RT/F = kT/e = 0.026$ V at 300 K. For reference in Figure 16.4 and

---

Chemists also define the so-called pK of water which is $pK = -\log K_W$. Because neutral water has $pH = pOH = 7$, the pH is always between zero and 14 and an equivalent measure is the *alkalinity*, $pOH = pK - pH = 14 - pH$.
[2]I should probably write this as

$$2H_3O^+(aq) + 2e^-(metal) \rightleftharpoons H_2(g) + 2H_2O$$

Chapter 17, note that the slope of $\mathcal{E}$ versus pH in both HER (16.2.6) and oxygen reduction (16.2.3) is $-2.303RT/F = -0.06$ V.

Table 14.1, Section 14.8, shows equilibrium exchange current densities and Tafel slope for the two important reduction reactions occurring next to an electrode, namely reactions (16.2.1) and (16.2.5). I need to emphasise that these are redox processes central to electrochemistry and corrosion but they differ from those we have studied up to now because they do not involve ions of the electrode metal. All the same, the inert electrode serves a number of functions. It may *catalyse* the reaction; it may offer a source or drain of electrons (generally a source as these are electronation reactions in the forward direction); it is clear from the table that whereas the standard equilibrium single potential is given by thermodynamics, namely zero for HER and $+0.4$ V for reduction of dissolved oxygen, the equilibrium exchange current density varies over eleven orders of magnitude for the HER. It seems that the excellent property of platinum as a catalyst is reflected in its large i_0 for these reactions.

16.3   Hydrogen Evolution Reaction

It is worthwhile to make a brief digression into some details of the HER. You might think that this very fundamental process, having been well studied over many years, is fully understood. On the face of it, it appears to be a simple piece of physics — the dissociation of water molecules to make hydrogen. In fact, it is vital that we *do* understand it since it must lie at the heart of the hydrogen economy of the future. However, surface physics is notoriously difficult, not to say irreproducible, in view of the difficulty of preparing identical surfaces in repeated experiments (you often hear this said in the world of transition metal oxides).

Figure 16.2 shows the current density associated with the HER as a function of the overpotential. Actually, as you will see in a moment when I describe the Evans diagram, it is conventional to plot the overpotential as a function of the logarithm (base 10) of the current density, in accord with the Tafel law (15.3.2). You can see that there is around an order of magnitude variation in overpotential needed to achieve, say, 1 mA cm^{-2} of current density *for the same reaction* but on different electrode surfaces. Indeed, there is also a variation

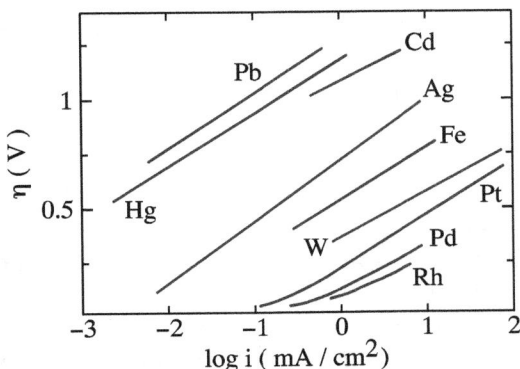

Fig. 16.2.   Overpotential of the hydrogen evolution reaction as a function of the log of the current density on a variety of metals. The variation between different metals is striking; if you draw a horizontal line at, say, 0.6 V, then you see there are about four and a half orders of magnitude variation in current density across the metals from Hg to Pt. Note that the Tafel slopes are all very similar and that apart from small deviations at low current density in Pt, Pd and Rh, the linear Tafel law is very closely adhered to over many decades of current density — it is clearly a very fundamental principle of electrodics. These are evidently Julius Tafel's own data.

Source: Adapted with permission from Electrochemistry Encyclopedia, knowledge.electrochem.org/encycl/art-t01-tafel.htm, copyright: The Electrochemical Society. (See also Figure 12.16 in Walter Moore, see Further Reading in Appendix D.)

depending on the state of the surface, and probably on the crystal face also.

In an acidic solution the HER is

$$2H_3O^+ + 2e^- \text{(in metal)} \rightleftharpoons H_2 + 2H_2O$$

As often in physical chemistry, the key question is, what is the rate limiting step? Walter J. Moore (see Further Reading in Appendix D) lists these as follows.

(i) Transport of the hydronium ion, H_3O^+, to the interphase. This will be governed by the principles of diffusion.

(ii) Reduction of the hydronium ion by transfer of an an electron from the metal. There are two proposed processes; they both begin with the *Volmer reaction*:

$$H_3O^+ + e^- \text{(in metal)} \rightleftharpoons H_{ads} + H_2O$$

Here, H_{ads} is a hydrogen atom (proton) *adsorbed* on the surface of the metal. Although simple in statement, this is complicated because surfaces are not perfect. Even if we have prepared a single crystal and the electrode is a flat, low-index surface, this will not happen on a surface terrace necessarily. We don't know whether this is more likely to occur at a ledge or kink site. Furthermore, the picture of a hydrogen atom standing proud with its bond like a flagpole emerging from a flat surface and waiting for the next event to occur is also problematic. Probably, surface diffusion is faster than the wait for the next proton to be discharged from the electrolyte so they can combine to H_2, and so the proton will skate over the surface to find a low energy adsorption site. Moreover, hydrogen is soluble in many metals and may well dive under the surface and nestle in one of the subsurface layers of metal atoms. Ultimately, the proton is a quantum object and these dynamic processes will be assisted by tunnelling. All the same, it is assumed that this is followed by one of two alternatives.

(1) The *Heyrovsky reaction*

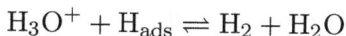

$$H_3O^+ + H_{ads} \rightleftharpoons H_2 + H_2O$$

In the Heyrovsky reaction, an *already existing* adsorbed hydrogen following a Volmer reaction combines with a proton belonging to a second hydronium to make a molecule of hydrogen.

(2) The alternative is that two adsorbed hydrogens appear as a result of two successive Volmer reactions closely spaced on the metal surface to make a hydrogen molecule. This is called the *Tafel reaction*

$$2H_{ads} \rightleftharpoons H_2$$

This is a surface reaction that presumably involves some surface diffusion of the two protons. When they meet, their s-orbitals will overlap and form a bonding orbital which stabilises the H_2 molecule. The strength of the H–H bond must depend on the position of the hydrogen s-orbitals relative to the Fermi level of the metal. If the Fermi level is higher, then both bonding and anti-bonding orbitals will be occupied and all other things being equal the bond strength will be zero.

This presumably means that the Volmer–Tafel mechanism will occur on non-transition metal surfaces. It is notable that in Figure 16.2 these metals, lead, mercury and cadmium, are distinct from the transition metals. In addition, the hydrogen evolution reaction occurs at a much lower overpotential on platinum, palladium and rhodium which are well-known catalysts. They are presumably catalysing the dissociation of water.

(iii) The third possible rate limiting process is the *desorption* of the hydrogen molecule, and this will depend on how tightly it is bonded to the metal.

(iv) Transport of the hydrogen through the electrolyte, either by diffusion or gas bubble formation, may be rate limiting.

Under normal conditions, it is assumed that one of the hydronium discharge processes in (ii) is rate limiting and the study of the Volmer–Tafel and Volmer–Heyrovsky processes is very intense and not yet concluded. A very interesting suggestion by Wilhelm *et al.* (*J. Phys. Chem. C*, **112**, 10814 (2008)) is that it is the *Zundel ion* that participates in the hydrogen adsorption. The Zundel ion is $H_5O_2^+$. From our research, it appears to us that the Zundel ion is two water molecules which are "sharing" a proton. It appears as a "flickering" event[3] in which two molecules are sharing an extra proton. Figure 16.3 shows the scene proposed by Wilhelm *et al.* The figure shows four "states" of the Zundel ion with its "shared" proton so it is actually an $H_5O_2^+$ ion that is handing one of its protons to the metal surface.

16.4 Corrosion Potential

Let us recap briefly. A piece of metal placed in a solution of its ions will acquire an electric potential difference across the interphase between its surface and the bulk solution. We usually refer to this

---

[3] *Flickering* is a term used in Monte Carlo and molecular dynamics computer simulations when a system is oscillating between two equivalent metastable states. In this case, the proton is jumping back and forth between the two water molecules, each in turn becoming briefly a hydronium ion. See https://tonypaxton.org/Water.html.

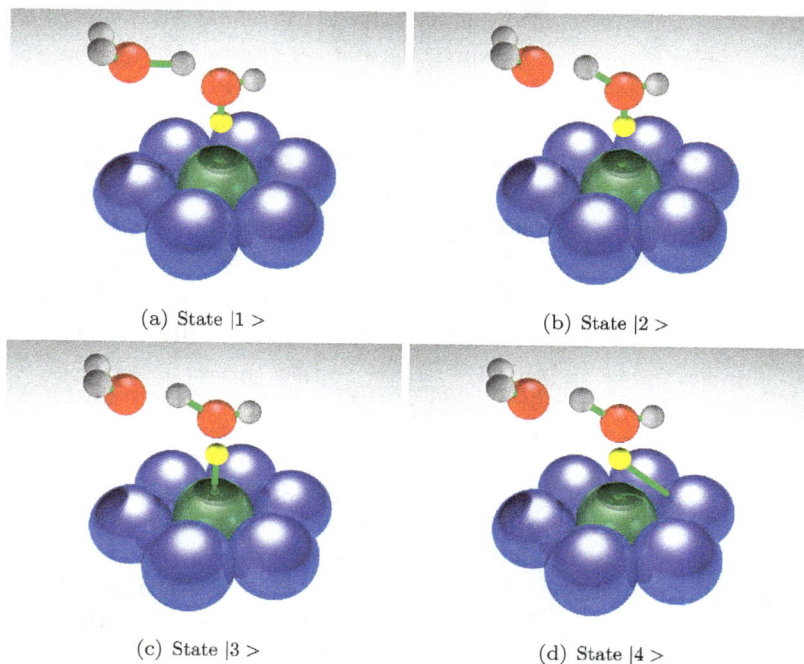

(a) State |1 >   (b) State |2 >

(c) State |3 >   (d) State |4 >

Fig. 16.3.   Atomic configurations proposed in a quantum chemistry study of the HER. Note that the hydronium ion at the centre, which donates a proton to the metal, is part of a Zundel ion. The metal surface is here represented as seven metal atoms, which is probably rather unrealistic since screening of charges by the metal's electron gas is excluded from consideration in the theory.
Source: Image taken with permission from Wilhelm *et al.*, *J. Phys. Chem. C*, **112**, 10814 (2008).

interphase loosely as the "double layer" (see Chapter 3). The numerical value of this potential difference can be calculated by reference to the electromotive force series (Table 9.3) and by use of the Nernst equation (if the activity of ions is not unimolal at 25°C). We have seen in Chapter 15 that this equilibrium potential difference may be modified by an *overpotential*, η, by connecting the metal electrically to an electron drain or source. We wrote

$$\Delta\phi = \Delta\phi_{eq} + \eta \tag{16.4.1}$$

and $\Delta\phi_{eq}$ is given by Equation (14.7.2). This potential difference is not measurable but we complete a circuit via high impedance voltmeter to an SHE by means of which we effectively subtract from both sides of (16.4.1) the potential difference of the SHE and we get

$$\mathcal{E} = \mathcal{E}_{eq} + \eta \qquad (16.4.2)$$

A simple question to ask is, will a metal dissolve in an acid solution at unit molality? The answer is yes if the metal's equilibrium electromotance is less (more negative) than that of the hydrogen evolution reaction which at unit activity is zero. So it's easy to find which metals will dissolve in acid at normal strength: it's those metals in the electrochemical series whose electromotance is negative — that is, all metals more base than copper. The next question is, how can this happen if the metal is electrically isolated? The answer, as I've revealed already, is that the metal divides itself into patches: at the anodes, the metal dissolves and at the cathodes, hydrogen is evolved (see Figure 16.1). But, wait! you say: the metal and the electrolyte are electrically conducting, how can there be different electric potentials at different parts of the metal? Or different parts of the electrolyte for that matter. Well, there can't. In addition to negotiations arriving at which patches will be anodes and which cathodes, there must also be agreement on a common $\Delta\phi$ across the interphase everywhere. So the metal chooses for itself what the potential will be everywhere. In effect, this is chosen so that when (16.4.2) is applied at the anode, the single potential is still greater than the equilibrium potential of the $M \rightleftharpoons M^{n+}$ equilibrium, while at the cathode, it is such that the potential is smaller (or more negative) than the equilibrium potential for the reduction reaction — either HER or reduction of dissolved oxygen. This negotiated potential difference is a so-called *mixed potential* and in corrosion science, it is called the *corrosion potential*, $\mathcal{E}_{cor}$. Our task is to find what it is and I will show a neat graphical method.

Meanwhile, let's do another example. Consider immersing a coupon of pure silver into neutral aerated water. Will it corrode? At first sight, the question is moot since the equilibrium potential of oxygen reduction is +0.8 volts (16.2.4) and so is the the value of $\mathcal{E}_M^\circ$ for Ag in the electrochemical series (Table 9.3). However, we have not accounted for the concentration of silver ions in the water. Let's suppose there is a trace amount so that their activity is 10^{-6}.

Then using the Nernst equation,

$$\mathcal{E}_{Ag} = \mathcal{E}_{Ag}^{\circ} + \frac{RT}{F} \ln h_{Ag^+} = +0.44 \, \text{V/SHE} \qquad (16.4.3)$$

so the silver is more base than the oxygen reduction on the metal surface (it doesn't matter that now the oxygen reduction is on silver and not platinum as in my previous calculation starting from (16.2.1) because in both cases the metal is inert since at the cathode patches the silver ions are not playing a role). I now calculate the "driving electromotance" of the combined reactions:

$$\mathcal{E} = 0.8 - 0.44 = 0.36 \, \text{V}$$

Combine this with the important equation from Section 9.3

$$\Delta G = -nF\mathcal{E} \qquad (9.3.6)$$

and we see that the free enthalpy change associated with corrosion of the silver coupon is

$$-0.36F = -34.7 \, \text{kJ mole}^{-1}$$

so indeed the free enthalpy decreases as the metal corrodes.

I hope this gives a clear picture of how metals corrode normally, without being connected to an external power supply as in all our previous examples. The key step is to ask whether an anodic de-electronation can occur simultaneously with a cathodic electronation (usually HER or dissolved oxygen reduction) and then work out the mixed potential and confirm that the overall combined processes at given concentrations (and temperature) will lead to an overall decrease in free enthalpy.

16.5   Evans Diagram

Right. The next question is, how fast does the metal corrode? Finally, I can put together the equilibrium thermodynamics of Section 14.7 and before, and the kinetics of Chapter 15 and I can show you a graphical construction that leads to a prediction of the corrosion potential and the rate of uniform corrosion.

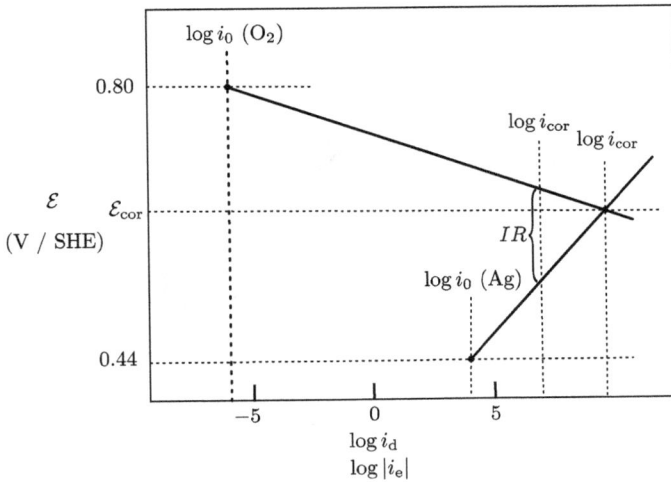

Fig. 16.4.   Evans diagram to illustrate the corrosion potential negotiated between anode and cathode regions (see Figure 16.1) of a silver coupon placed in pure water.

Return to the example in which I place a coupon of silver into neutral, aerated water. A corrosion cell as sketched in Figure 16.1 is set up. I have all the data I need in (16.2.4), (16.4.3) and Table 14.1, Section 14.8. I construct a plot of electromotance, $\mathcal{E}$, versus the logarithm of the current density as in Figure 16.4. I assume that the anode and cathode areas are equal — otherwise I must plot current, not current density. I mark off on the ordinate the equilibrium single electrode potentials of the metal de-electronation reaction (16.4.3), namely 0.44 V/SHE, and the oxygen reduction, 0.8 V/SHE (16.2.4). I mark off on the abscissa the logarithms of the corresponding equilibrium exchange current densities, i_0, taken from Table 14.1. I then draw straight lines, assuming the Tafel law (15.3.2) and using the Tafel slopes taken from Table 14.1. The Tafel line for the de-electronation, metal wastage, slopes upwards since the overpotential increases with increasing current density (15.3.1a) and conversely for the electronation (oxygen reduction). These are called the anodic and cathodic polarisation curves, respectively. Now you see how the metal negotiates its corrosion potential, $\mathcal{E}_{cor}$, and corrosion current density, i_{cor}. The electronation and de-electronation current densities must be equal to conserve charge (I am assuming equal areas)

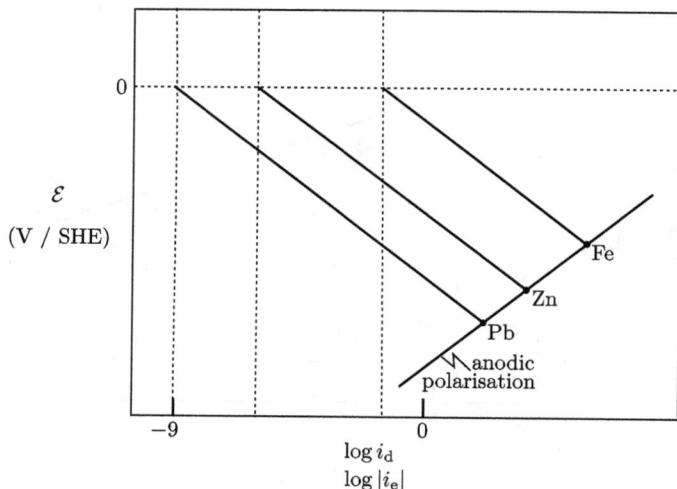

Fig. 16.5. Evans diagram to demonstrate the different corrosion potentials adopted by hydrogen reduction on different metal electrodes.

and so this must be found by a vertical line drawn on the diagram. If the electrolyte is perfectly conducting (as of course is the metal), then the corrosion current density must be at the intersection of the two Tafel lines as shown. In case the electrolyte is not perfectly conducting, there may be an "IR" drop in the electrolyte between anode and cathode, in which case, as shown, i_{cor} must be common to both anode and cathode, but now the interphase potentials are not common to anode and cathode. But in the case of zero resistance, the intersection of the two Tafel lines indicates not only i_{cor} but also the common corrosion potential, $\mathcal{E}_{cor}$, as I described it in Section 13.1.

Figure 16.4 is called an "Evans diagram" after the pioneering corrosion scientist, Ulick R. Evans (1889–1980). There are a million uses of the Evans diagram to illustrate corrosion processes: crevice attack, differential aeration and countless others. Here is a crude example which is nonetheless insightful. In Figure 16.5, I show an anodic polarisation curve for a generic metal assuming it's the same for, say, lead, zinc and iron (although you see from Table 14.1 that they actually are very different). I then show cathodic polarisation curves for the hydrogen evolution reaction on these three metals taking account that they all have the same Tafel slope, but very different exchange current densities, taken from Table 14.1. This is also very

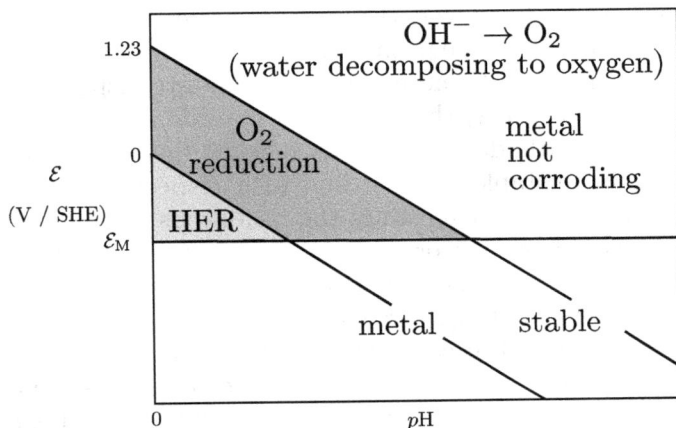

Fig. 16.6.   Electromotance plotted against pH showing regimes of equilibrium of competing electrochemical processes.

rough and ready since these are i_0 values in water and the HER is only the relevant cathode reaction in acids — exchange current densities in normal strength sulphuric acid, say, are typically about ten times what they are in pure water. All the same, the Evans diagram illustrates a key fact that the rate of corrosion of these three metals will vary by several orders of magnitude in the same strength acid. This is the basis of *galvanisation* in which a coating of zinc is used to protect steel. Not only is the corrosion rate greatly reduced but so is the rate of hydrogen evolution, and this provides a means of protection of the steel from hydrogen embrittlement.

In Figure 16.4, I treat a case in which the cathode reaction as in Figure 16.1 is oxygen reduction, while in Figure 16.5, I expect it to be hydrogen evolution because conditions in the electrolyte are acidic. Can I decide in general what will be the cathode process in a given corrosion situation? In Figure 16.6, I plot electromotance against pH, not current density, so the single electrode potential on the SHE scale, $\mathcal{E}_M$, appears as a thick horizontal line as it does not depend on pH. At any electrode potential below this line, the metal is immune from corrosion. Now I plot the equilibrium potentials of the oxygen reduction and HER which do depend on pH, as expressed in Equations (16.2.3) and (16.2.6). At electrode potentials *above* the oxygen reduction equilibrium in Figure 16.6, the reaction (16.2.1) is driven to the left and indeed water decomposes to oxygen at such a large positive potential. Moreover, the metal may not corrode because its electrode potential is more positive than either possible electronation reactions so there can be no mixed potential established. Below this line but above $\mathcal{E} = \mathcal{E}_M$, the metal can corrode and oxygen can be reduced at the cathode. At potentials below the HER sloping line, water is again unstable — this time with respect to hydrogen gas (you can see how this determines the conditions for the electrolysis of water to either oxygen or hydrogen gas, essential to the "hydrogen economy"). In the shaded triangle, the electrode potential is sufficiently base that the hydrogen evolution reaction may displace oxygen reduction at the cathode; this will happen in acidic conditions (low pH) and in the absence of dissolved oxygen. Therefore, you will often read a statement such as to the effect that a metal at an electrode potential lower than the potential for oxygen reduction or hydrogen evolution will become oxidised. The cathode reaction in the lemon lamp of Chapter 1 is the HER because lemon juice is acidic and doesn't contain much dissolved oxygen.

Chapter 17

Surface Film and Pourbaix Diagram

I have called the experiment of placing a coupon of pure metal into water, or a solution of its ions, a *thought experiment* partly because there is no pure metal, but much more significantly because all metals, except gold, are likely to have surface films of oxide or hydroxide and so the pure metal electrolyte interface is only achieved under certain conditions as we now discover. You are particularly aware that aluminium, titanium and chromium, among many other metals, are protected from corrosion in damp air not because they are noble (in fact, they are base — see Table 9.3, Section 9.5) but because they are protected by a film of oxide, possibly only a few nanometres thick, which coheres tenaciously to the surface. Other base metals, such as iron, are not protected in this way because their oxides, possibly due to a mismatch in atomic volume, flake off and expose new metal to be oxidised. Alloying of steel with chromium led to the fantastically important discovery of stainless steel by Harry Brearley. If you can discover stainless magnesium, you will make your fortune.

17.1   Passivation

The benefits of the formation of an oxide or hydroxide film can be sketched out in an Evans diagram. The details are quite complicated but I can give a qualitative description.

In reference to Figure 17.1, begin at the lower left of the anodic polarisation curve of a metal, such as stainless steel or aluminium. As the potential is raised, the current density increases with a normal

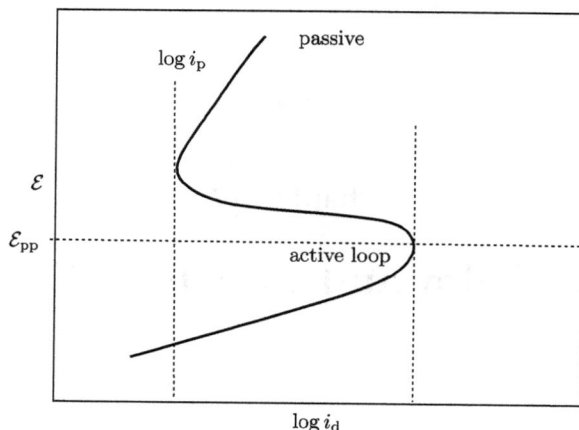

Fig. 17.1. Evans diagram illustrating passivation. The polarisation curve does not continue to rise as a function of current density; instead at $\mathcal{E} = \mathcal{E}_{pp}$, the peak passivation potential, a protective oxide film will form so that as the overpotential is raised the current density actually drops. The film may break down at a current density, i_p, after which the polarisation curve continues, typically having a greater Tafel slope than before passivation initiated. At overpotentials below $\mathcal{E} = \mathcal{E}_{pp}$, we are said to be in the *active region*, above that the system is *passive* and *transpassive* at overpotentials above the *passive loop*.

Tafel slope just as in Figure 16.4. As the metal oxidises, M^{n+} ions may detach and migrate through the double layer as I have already described, or under the right conditions, the M^{n+} ions may combine with oxygen or hydroxide ions from the solution to form a solid oxide or hydroxide film. This may become a tenacious ceramic coating a few ones or tens of nanometres thick. Clearly, this will greatly affect the kinetics and we now must regard the film and double layer acting in parallel, each with its de-electronation rate coefficient and symmetry factor. In fact, transport of electrons and cations across the film is a matter of electrical conduction or quantum tunnelling of electrons and vacancy mediated solid state diffusion of, indeed, anions as well as cations. Without doing the maths, the result is that once the electrode potential is sufficiently raised to what is called the peak passivation potential, $\mathcal{E}_{pp}$, at some maximum current density as indicated in Figure 17.1, the kinetics become a great deal more sluggish and the current density drops to a value i_p, called the passivation current density, and this may represent several orders of magnitude reduction in corrosion rate. Thereafter, as may be demonstrated by calculation (see J. M. West in Further Reading, Appendix D), the Tafel slope is increased by as much as a factor two so that even if the electrode potential is further raised, the increase in current density is much less rapid. For example, in an 18–8 stainless steel in de-aerated sulphuric acid at 25°C, the peak passivation potential is −0.2 V/SHE, and the passivation current density is as small as 0.006 Am^{-2}, which is not much if you remember that equates to a similar number of mm per year metal wastage.

17.2   Pourbaix Diagram

Figure 16.6 is a simple kind of "Pourbaix diagram". This is an equilibrium diagram, or phase diagram if you like, indicating the state of equilibrium of an electrode–electrolyte system in the space of electromotance (single potential) and acidity, pH. It allows you to predict whether a metal will corrode and what will be the corrosion products for a given state represented by an electrode potential (on the SHE scale) and the pH of the solution — all other things being equal, of course, including concentrations of ions in solution and temperature. In what follows, I show how the simplest Pourbaix diagram is constructed from thermodynamic data and then take a short tour of the Pourbaix diagram for iron.

17.2.1   *Separation into fields in the Pourbaix diagram*

Assume for simplicity a divalent M^{n+} ion, so $n = 2$, which allows the formation of a simple stoichiometric oxide, MO. Then there are two possible electronation reactions. The first is

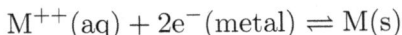

$$M^{++}(aq) + 2e^-(metal) \rightleftharpoons M(s)$$

for which the equilibrium potential is

$$\mathcal{E}_{M^{++}} = \mathcal{E}^\circ_{M^{++}} + \frac{RT}{2F} \ln h_{M^{++}} \qquad (17.2.1)$$

and you could look up $\mathcal{E}^\circ_{M^{++}}$ in the electrochemical series, Table 9.3. The second possible reaction involves the formation of the oxide film

$$MO(s) + 2H^+ + 2e^-(metal) \rightleftharpoons M(s) + H_2O(l)$$

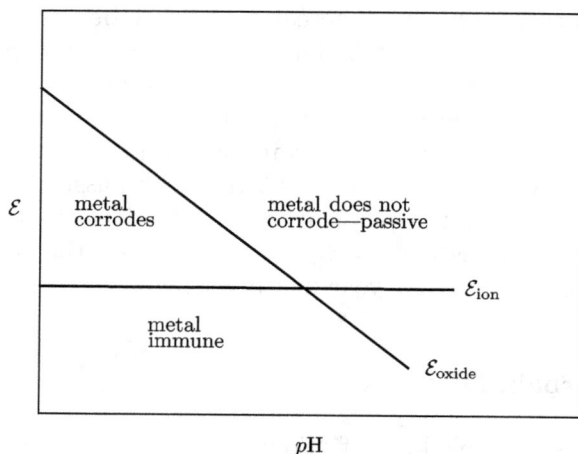

Fig. 17.2.   Schematic of regions in the pH $\mathcal{E}$ space.

for which the equilibrium potential is

$$\mathcal{E}_{MO} = \mathcal{E}_{MO}^{\circ} + \frac{RT}{F} \ln h_{H+}$$

$$= \mathcal{E}_{MO}^{\circ} - 0.06 p\text{H} \quad \text{V/SHE at } 25°\text{C} \tag{17.2.2}$$

I can plot on a graph of $\mathcal{E}$ against pH, the lines (17.2.1) and (17.2.2). This is sketched in Figure 17.2. Thermodynamics indicates that for conditions of pH and electrode potential to the right of the sloping line, the metal is unstable with respect to its solid oxide. This means that as long as the oxide is able to form a tenacious and continuous surface layer, then we will be in the passive region of Figure 17.1 and the metal will not corrode. If the electrode potential is smaller (more negative) than $\mathcal{E}_{ion}$ ($\mathcal{E}_{M++}$), the metal is at a lower potential than its equilibrium $\Delta\phi_{eq}$ and will not corrode as long as we are to the left of the sloping line so that it is stable with respect to its oxide. This region of behaviour is sometimes called "immunity". The third domain of behaviour is the closed triangle in which a metal is not passive nor immune and will corrode.

You can imagine superimposing the oxide sloping line onto the Pourbaix diagram in Figure 16.6, which will add additional data since that previous diagram was drawn without taking into consideration the possibility of passivity. You then have a full picture of

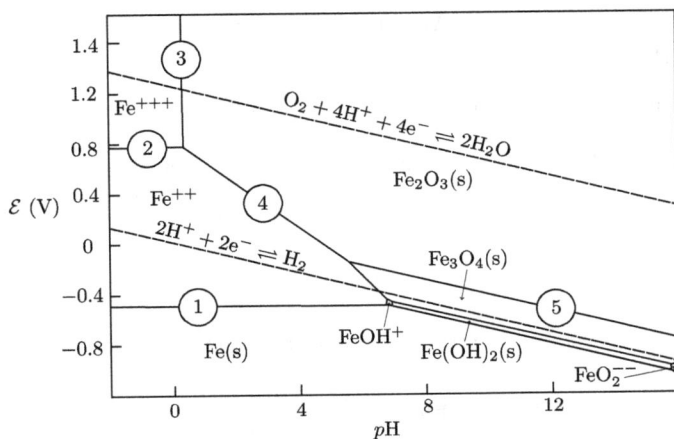

Fig. 17.3. Pourbaix diagram for iron with phase boundaries labelled in accord with some of the reactions given in the text.
Source: Shared under a CC BY-SA 3.0 license, authored by PennState Eberly College of Science, https://science.psu.edu/chem/undergrad/courses.

the prevailing conditions that will determine corrosion, passivity and immunity, and if corrosion is indicated, what is the likely accompanying cathode reaction (see Figure 16.1).

17.3 Pourbaix Diagram for Iron

Figure 17.3 provides us with sufficient pedagogical material for a brief tutorial. Reading a Pourbaix diagram is not unlike reading a binary temperature composition phase diagram in which for a given T and c the thermodyamically stable phase or phase mixture is indicated by the field in which the T and c combination falls. The phase boundaries in a Pourbaix diagram may be vertical, horizontal or straight with a constant slope.

The Pourbaix diagram for iron is sufficiently complicated to include all three types of phase boundaries. It is complicated because iron can exist in at least two forms of oxide, Fe_2O_3 and Fe_3O_4. The Pourbaix diagram, Figure 17.3, labels five of these lines and the equilibria whose right- and left-hand sides are separated by such phase boundaries are described now. In these calculations, I will assume that Fe^{++} and Fe^{+++} ions in solution are at Henrian activity (molal concentration) of 10^{-6}.[1] Also recall that the activities of pure substances (including water, as it's very nearly pure) are one.

(1) Reduction (electronation) of Fe^{++} to metal iron:

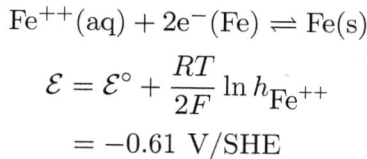

$$Fe^{++}(aq) + 2e^-(Fe) \rightleftharpoons Fe(s)$$

$$\mathcal{E} = \mathcal{E}^\circ + \frac{RT}{2F} \ln h_{Fe^{++}}$$

$$= -0.61 \text{ V/SHE}$$

This reaction's electromotance is independent of pH so appears as a horizontal line.

(2) Reduction (electronation) of Fe^{+++} to Fe^{++} (also independent of pH, so horizontal):

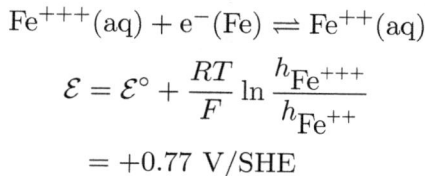

$$Fe^{+++}(aq) + e^-(Fe) \rightleftharpoons Fe^{++}(aq)$$

$$\mathcal{E} = \mathcal{E}^\circ + \frac{RT}{F} \ln \frac{h_{Fe^{+++}}}{h_{Fe^{++}}}$$

$$= +0.77 \text{ V/SHE}$$

(3) Equilibrium of Fe_2O_3 and Fe^{+++}:

$$Fe_2O_3(s) + 6H^+(aq) \rightleftharpoons 2Fe^{+++}(aq) + 3H_2O(l)$$

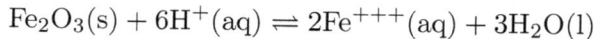

In this reaction, there is no charge transfer to or from the electrode, no reduction or oxidation, hence the boundary is independent of $\mathcal{E}$ and appears at a fixed pH.

(4) Reduction (electronation) of Fe_2O_3 (slope -0.177 V/pH):

$$Fe_2O_3(s) + 6H^+(aq) + 2e^-(Fe) \rightleftharpoons 2Fe^{++}(aq) + 3H_2O(l)$$

$$\mathcal{E} = \mathcal{E}^\circ + \frac{RT}{2F} \ln \frac{h_{H^+}^6}{h_{Fe^{++}}^2}$$

$$= +0.73 - 0.117pH - 0.06 \log 10^{-6}$$

$$= +1.08 - 0.177pH \text{ V/SHE}$$

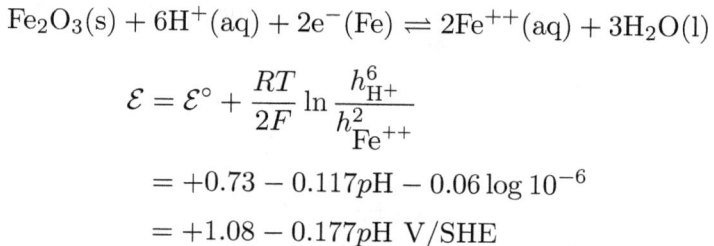

[1]There is no fundamental reason for this assumption, but it's commonly used because it works (see also Problem 17.2).

(5) Reduction (electronation) of Fe_2O_3 to Fe_3O_4 (slope -0.06 V/pH):

$$3Fe_2O_3(s) + 2H^+(aq) + 2e^-(Fe) \rightleftharpoons 2Fe_3O_4(s) + H_2O(l)$$

$$\mathcal{E} = \mathcal{E}^\circ + \frac{RT}{2F} \ln h_{H^+}^2$$

$$= +0.22 - 0.06pH \text{ V/SHE}$$

Also shown in Figure 17.3 are the oxygen reduction and hydrogen evolution lines which I showed in Figure 16.6. So the Pourbaix diagram for iron is complete and you can read off the domains of behaviour. The passive field is now strangely shaped because of the different stoichiometries of iron oxides and the iron hydroxides that I have not discussed. Above the oxygen reduction line, there may be neither corrosion nor passivation since the water is unstable with respect to gaseous oxygen. The corroding field is bounded by the lines 1, 2 and 4, and the field at the bottom is the region of immunity. This is summed up in Figure 17.4.

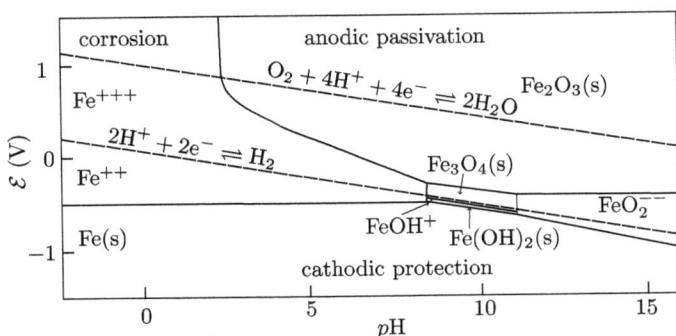

Fig. 17.4. Pourbaix diagram for iron showing regions of equilibrium.
Source: Shared under a CC BY-SA 3.0 license, authored by PennState Eberly College of Science, https://science.psu.edu/chem/undergrad/courses.

Chapter 18

The Evils of Chloride

The engineer can deal with uniform corrosion and the benefits of passivation are obvious. The reason as I stated in the beginning of Chapter 13 that we cannot leave it at that is that almost all corrosion failures are a consequence of localised attack at crevices, dissimilar metal joints, regions of reduced aeration and so on. But *physics* cannot take us there and so we reach the limits of our scope except for some brief remarks which follow. While what we have learned has great value in many branches of modern science in particular batteries and fuel cells for e-mobility, I have avoided those topics and concentrated in these last sections on corrosion. My hope is that this text has equipped the reader with the tools to proceed into these areas, and of course, the climate emergency calls upon us rapidly to develop better batteries and fuel cells.

When I was an undergraduate, I was taught by John West, and I also took a job in the long vacations at the Sheffield University Metals Advisory Centre (SUMAC) in which Graeme Davies was Chairman and Tony Entwisle was Managing Director, not to mention the colourful John Devine. Our business was mostly failure investigation with a bit of alloy design. I remember two corrosion problems and in one instance, I suggested to Tony that we ask John West for an opinion. Tony didn't think that was called for and that's when he issued his wonderful remark: you only need to know two things in a corrosion investigation; anodes are positive and they corrode; and find where the chloride came from. He was right and I found it! It is of course common in seawater and greatly complicates marine corrosion.

Of hydrogen embrittlement, Dave Rugg of Rolls-Royce famously said, *hydrogen is intelligent and it's evil*. Almost the same can be said of the chloride ion. Just the salt from handling after eating fish and chips can lead to an aerospace disaster. Chloride plays at least two roles in the devastation of engineering structures. One is that it attacks passive films. So Pourbaix diagrams must be re-drawn in sea water, with the passive regions shrinking dramatically.

Second is its role in pitting corrosion. Uniform corrosion can be handled, but if corrosion becomes localised, say due to patches of metal in regions of electrolyte depleted of dissolved oxygen, then if these regions have a small area, the anode current density greatly exceeds the cathode current density and the metal is effectively drilled by anodic dissolution. Once a pit forms, the process is self-sustaining and irrecoverable. The electrolyte in the pit constitutes a "micro climate". Detailed thermodynamic calculations reveal the effect of chloride ions in the pit is to lower the pH, to raise the partial pressure of hydrogen and to lower the equilibrium potential. Drilling will continue until a plate or pipe is perforated and it fails.

Appendix A

Outline of the Thermodynamics of Metals and Solutions

A.1 Closed Systems

A.1.1 *State functions*

> All thermodynamic thinking begins with a definition of the portion of the universe under study, i.e., the system.
>
> John O'M. Bockris

I will not start right at the beginning but I will try and introduce what we will need in thermodynamics using just what you have already learned about the first and second laws.

The first law states that for a "closed" system, changes in the heat content and the amount of work done *on* the system amount to a change in *internal energy*[1] written as

$$dU = \delta q + \delta w \qquad (A1.1)$$

---

[1]From a mechanical point of view, we often state that the total energy of a body, say a piece of Cu–Ni alloy, is equal to its potential energy, E_{pot}, if it's in a gravitational field, say, plus its kinetic energy, E_{kin}, if it's moving relative to some inertial frame. This neglects the *internal* state of the body, that is to say, what its atoms are doing and whether any diffusion, phase transformations or chemical reactions are taking place, whether it is expanding or getting hotter or colder. So in thermodynamics as opposed to mechanics, we express the total energy as $E_{tot} = E_{pot} + E_{kin} + U$ and since thermodynamics concerns only *changes* in total energy, we consider systems for which the potential and kinetic energies are unchanging, for example, that the body is not in motion, and focus purely on the internal energy.

If this is combined with the second law, then this becomes

$$dU = T dS - p dV \tag{A1.2}$$

This is true as long as the only work done involves a change in volume, V, against an external pressure, p. The absolute temperature is T and S is the entropy. A *closed* system is one that does not exchange matter with its surroundings (although it may transfer heat). We will come on to *open* systems in a while.

Before that, we note that (A1.2) is a complete statement of the combined first and second laws for a closed system if only pV-work is done.[2] However, it may not be the most convenient. For example, we may like to simplify things by doing experiments so that one of the two terms on the right-hand side of (A1.1) is zero. So if we want to know just the result of the system doing some work, then working with the internal energy as a function of entropy and volume, $U(S, V)$, requires us to do experiments adiabatically, that is, under the condition $\delta q = 0$, and this is not very easy.

So without making any additional postulates we devise a number of *auxiliary functions*. These are

$$\text{enthalpy,} \quad H = U + pV$$
$$\text{Helmholtz function,} \quad F = U - TS$$
$$\text{Gibbs function,} \quad G = H - TS$$

G is also known as the *free enthalpy* or *Gibbs free energy*; F is also known as the *free energy* or *Helmholtz free energy*. We are most interested in the Gibbs function so I'll show you how we get it from Equation (A1.2). We do it in two steps. First, we have that U is a function of S and V as indicated by (A1.2). Can we find an auxiliary function that depends on T and V instead? What is the relation between T and S? From (A1.2), we have

$$T = \left(\frac{\partial U}{\partial S} \right)_V$$

---

[2]We have to extend this in cases where there is surface tension, non-uniform stress, changes in electric or magnetic state and so on. For example, work done on a body by a non-hydrostatic stress is $\sigma_{ij} d\varepsilon_{ij}$, or work done on a body by increasing the area, A, of an internal interface of energy γ per unit area (say between the ferrite and austenite phases in steel) is γdA. Electrical work on a linear dielectric is $\mathbf{E} \cdot d(\epsilon \mathbf{E})$.

We say that S and T are "conjugate variables", and we invent the new function

$$F = U - TS \qquad (A1.3)$$

that is, the first function, U, take away the product of the two conjugate variables (the one we are trying to get rid of and replace with the other). Now by taking the total derivative (that means, asking what a change in F results in),

$$\mathrm{d}F = \mathrm{d}U - \mathrm{d}(TS) = \mathrm{d}U - T\mathrm{d}S - S\mathrm{d}T$$
$$= -S\mathrm{d}T - p\mathrm{d}V \qquad (A1.4)$$

using (A1.2). Now we have a function F which depends on T and V rather than U which depends on S and V so that we can more readily interpret experiments done at constant volume or constant temperature. Suppose we are working at constant pressure rather than constant volume, then we need yet another auxiliary function, which we obtain from F by replacing the variable V with its conjugate $-p$. Indeed, we know from (A1.4) that

$$p = - \left(\frac{\partial F}{\partial V} \right)_T$$

so $-p$ and V are conjugate variables, meaning that I can find a function that depends on p and T by writing down

$$G = F + pV$$
$$(= H - TS) \qquad (A1.5)$$

(the original function take away the product of the two conjugate variables[3]). Then similar to before,

$$\mathrm{d}G = \mathrm{d}F + p\mathrm{d}V + V\mathrm{d}p$$
$$= \mathrm{d}U - T\mathrm{d}S - S\mathrm{d}T + p\mathrm{d}V + V\mathrm{d}p$$
$$= T\mathrm{d}S - p\mathrm{d}V - T\mathrm{d}S - S\mathrm{d}T + p\mathrm{d}V + V\mathrm{d}p$$
$$= -S\mathrm{d}T + V\mathrm{d}p \qquad (A1.6)$$

and is thereby shown to be the required auxiliary function of T and p.

---

[3]In mathematics, this is called a *Legendre transformation*. Note that for a pair of conjugate variables, one is intensive and the other extensive.

A.1.2   *Conditions for equilibrium in closed systems*

According to the second law, in an *isolated system* (one that can exchange neither matter nor energy with the environment), if any change takes place, it must be such that the entropy increases or remains constant:[4]

$$dS \geq 0$$

Strictly in thermodynamics, when we write

$$d(\text{something})$$

we really mean

$$\frac{d(\text{something})}{dt}$$

but we accept that the "arrow of time" always runs in the positive direction even if we don't quite understand why and we leave out the dt denominator. If the isolated system is made up of one or more parts, then we have

$$\sum_m dS_m \geq 0$$

each part being labelled by a subscript m. So some part may suffer a decrease in entropy as long as the total entropy does not decrease. So we can think of a closed system as an isolated system made up of two parts — the part we are interested in (say, a block of alloy) and a large reservoir of heat at a fixed temperature T. In this way, we can keep our body at constant temperature if it is kept in contact with the reservoir. So if the entropy of our body is S and that of the reservoir is S_r, then, for a natural process,

$$dS + dS_r \geq 0 \tag{A1.7}$$

---

[4]We believe that the universe is an isolated system, hence the famous statement of Rudolf Clausius, *Die Energie der Welt ist konstant. Die Entropie der Welt strebt einem Maximum zu*: The energy of the world is constant. The entropy of the world is striving to a maximum. This will of course lead to the "heat death" of the universe, but don't worry the sun will have grown into a red giant and swallowed the earth by then.

Now for an infinitesimal change in heat content of the body, δq, at a temperature T, the change in entropy is $\delta q / T$;[5] this means that the change in entropy of the reservoir where the heat came from is $dS_r = -\delta q / T$ (because that amount of heat has been taken out of the reservoir at constant temperature T) and so putting this into (A1.7), the second law has it that

$$dS - \frac{\delta q}{T} \geq 0$$

Now using (A1.2), we have

$$dS - \frac{dU - \delta w}{T} \geq 0$$

I multiply through by T, rearrange and note that since this change is at constant temperature, $d(U - TS) = dU - TdS$ and I get

$$d(U - TS) \leq \delta w \qquad (A1.8)$$

By comparison with (A1.3) we have $dF \leq \delta w$ so that if the body does work ($\delta w < 0$), then the Helmholtz function may increase, but otherwise and especially if no work is done either by or on the body, then the Helmholtz function can only decrease during a natural change, or remain constant. Thus when all changes have happened and the body is in equilibrium, then the Helmholtz free energy is at a minimum. This is a condition for equilibrium.

I can add $d(pV)$ to both sides of Equation (A1.8) and I get

$$d(U + pV - TS) = dG \leq \delta w + d(pV)$$

If as well as working at constant temperature, my body is maintained under a constant external pressure, p, then $d(pV) = pdV$ and so if I define $\delta w' = \delta w + pdV$ as the work done not including work done, pdV, by the body against the external pressure,[6] then the condition

---

[5]You can think of this as the *definition* of entropy if you like: if an infinitesimal quantity of heat, δq, is added to a body reversibly and at constant temperature T, then the body's entropy increases by the amount $\delta q / T$.

[6]In this text, δw is the work done on the body, not the work done by the body. The latter convention is commonly used by engineers who are interested in, say, an internal combustion engine for which pdV is the work done by the explosion of a fuel moving a piston against the external pressure p. So for us, if the body does work against the external pressure, then the work done *on* the body is $-pdV$. So $\delta w' = \delta w - (-pdV) = \delta w + pdV$.

for equilibrium is that the Gibbs free energy of a body at constant temperature and pressure is a minimum if no work other than that either *against* or *by* the external pressure is done either on or by the body,

$$dG \leq \delta w' \quad \text{in a closed system}$$

If no such work is done ($\delta w' = 0$), then $dG \leq 0$ in which case the Gibbs free energy can only decrease or remain constant. This means that when all changes have happened that are going to happen, the system is in equilibrium and G is a minimum, $dG = 0$. This is the condition for equilibrium of a closed system at constant temperature and pressure.

In metallurgy when we deal with solids, the amount of work done, say, by the thermal expansion of a body against atmospheric pressure is so tiny as to be negligible, and we can regard the Helmholtz and Gibbs functions to be interchangeable and both minimal at equilibrium, but since we usually work at constant pressure and not constant volume, the focus is always on the Gibbs free energy.

A.2 Open Systems, Chemical Potential

An "open" system, such as the lump of metal that we have been thinking about, can exchange both energy and matter with its surroundings. Under these circumstances, we must modify our statement of the combined first and second laws (A1.2). We have to ask about the chemical composition of our body and to identify how many different "components" it is made up from. It is sufficient for our purposes to identify each chemical element as one of the components so that, for example, a piece of Cu–Ni alloy has two components. In addition to components, the body may be divided into phases. For example, a piece of Fe–C alloy at equilibrium within the $\alpha + \gamma$ field of the Fe–C phase diagram may have a certain volume fraction of ferrite, having a very small concentration of carbon and a certain volume fraction of austenite with a much larger concentration of dissolved carbon. These two phases will be intimately in contact sharing one or more interfaces which divide the body into its phases. For now, we consider a homogeneous, single phase body having N components and we will use a subscript i to label these in our mathematical formulas. The internal energy is now no longer a function of only the two variables S and V; it is also a function of the number of moles,[7] n_i, of each component that currently make up the body:

$$U = U(S, V, n_1, n_2 \ldots n_N)$$

The total differential of the internal energy is, now,

$$dU = \left(\frac{\partial U}{\partial S}\right)_{V,n_i} dS + \left(\frac{\partial U}{\partial V}\right)_{S,n_i} dV + \sum_{i=1}^{N} \left(\frac{\partial U}{\partial n_i}\right)_{S,V,n_{j \neq i}} dn_i \quad (A2.1)$$

In the case of constant composition, this is obviously still valid, and so from (A1.2),

$$\left(\frac{\partial U}{\partial S}\right)_{V,n_i} = T \qquad (A2.2)$$

and

$$\left(\frac{\partial U}{\partial V}\right)_{S,n_i} = -p \qquad (A2.3)$$

We now define the *chemical potential* of component i as the term under the summation sign in (A2.1):

$$\mu_i = \left(\frac{\partial U}{\partial n_i}\right)_{S,V,n_{j \neq i}} \qquad (A2.4)$$

and here the partial derivative is taken of U with respect to the number of moles of component i while keeping all other variables constant, namely the volume and entropy and the numbers of moles of all the other components. You might ask, how can I do this in practice? The answer is to imagine that to the body in question, you

---

[7]If you're a physicist, you tend to think in terms of the number of particles or number of atoms of each component, but since we deal with macroscopically sized bodies, in metallurgy and chemistry, we use the number of moles to keep the numbers of reasonable size. A mole is nothing other than an Avogadro number of objects and the Avogadro number is $L = 6.023 \times 10^{23}$. Physicists use the Boltzmann constant, k, while we use the gas constant $R = Lk$.

take an infinitesimal number of moles of component i (from a reservoir in which its chemical potential has some standard value — more on that later) and add it to the body. While the body's volume necessarily changes, you readjust that by application of an infinitesimal increase in pressure. Since the added quantity of matter may bring with it some heat, to ensure that the entropy doesn't change in this process, it is necessary then to remove that heat, say by placing it in contact with a heat bath at the appropriate temperature.

Maybe you'd like to think of the chemical potential as the reversible work done in bringing an infinitesimal amount of matter from a reservoir to the body in question. Sometimes I like the analogy with electrostatics; the *electric potential* is the work done in taking a positive test charge from infinity (the reservoir) to a place in space where there is an electric field — that is, the influence of other charges. Loosely speaking, if the electric potential is large and negative, then positive charges are attracted to that place and the work done in bringing the test charge from infinity is negative; conversely, if the electric potential $\phi(\mathbf{r})$ is positive at position $\mathbf{r}$, then I need to do work to bring in extra positive charges. In this vein, if the chemical potential of, say, carbon in Fe–C is large and positive, then I need to expend energy to increase its concentration. If the concentration, or rather the chemical potential, of C in Fe varies from place to place, then if the carbon is mobile, it will diffuse from regions of high chemical potential to regions of low chemical potential. This is rather obvious if we equate chemical potential with concentration. Later, we'll see how these two are actually related. Whereas in standard diffusion theory, you are told that the carbon will travel down a concentration gradient; to be properly precise, you should say that it diffuses down the gradient in *chemical potential*.

In view of Equations (A2.1)–(A2.4), we have for the total differential of the internal energy of a body:

$$dU = TdS - pdV + \sum_{i=1}^{N} \mu_i dn_i \qquad (A2.5)$$

which is the modification of the combined first and second laws (A1.2) for the case of an open system.

Now let me begin this argument again, but this time starting with the modification of the Gibbs function (A1.5) for an open system. The Gibbs free energy is a function of temperature, pressure and the numbers of moles of each of the components,

$$G = G(T, p, n_1, n_2 \ldots n_N)$$

and so its total differential is

$$dG = \left(\frac{\partial G}{\partial T}\right)_{p,n_i} dT + \left(\frac{\partial G}{\partial p}\right)_{T,n_i} dp + \sum_{i=1}^{N} \left(\frac{\partial G}{\partial n_i}\right)_{T,p,n_{j\neq i}} dn_i \quad (A2.6)$$

We now define the *chemical potential* of component i as

$$\mu_i = \left(\frac{\partial G}{\partial n_i}\right)_{T,p,n_{j\neq i}} \quad (A2.7)$$

and (A2.6) being still valid in the special case of all the dn_i being zero (that is, fixed composition) Equation (A1.6) gives

$$\left(\frac{\partial G}{\partial T}\right)_{p,n_i} = -S; \quad \left(\frac{\partial G}{\partial p}\right)_{T,n_i} = V \quad (A2.8)$$

which when put into (A2.6) results in

$$dG = -SdT + Vdp + \sum_{i=1}^{N} \mu_i dn_i \quad (A2.9)$$

with μ_i defined by Equation (A2.7). This is the modification of (A1.6) for an open system. Have I now made two different definitions of the chemical potential? That would be a real mess. Well, luckily, no! If I add to both sides of (A2.5) the quantity $d(pV - TS)$, then this equation is transformed into Equation (A2.9) by virtue of the fact that $G = U + pV - TS$.

We use Equation (A2.7) as our expression for the chemical potential (numerically, it is identical to (A2.4), but we are concerned in metallurgy with the Gibbs function whose independent variables are p and T which are easily controlled — it's not easy to measure let alone control the entropy!). In the case of there being a single component, the chemical potential is the "partial free energy per mole". Usually, for all extensive state functions, we define a partial amount, being the amount per mole. We use lower case for these. In this way, we have for a single component homogeneous body,

$$\text{partial molar volume,} \quad v = V/n$$
$$\text{partial molar enthalpy,} \quad h = H/n$$
$$\text{partial molar entropy,} \quad s = S/n$$
$$\text{partial molar free energy,} \quad f = F/n$$
$$\text{partial molar free enthalpy,} \quad g = G/n$$

Equation (A2.7) is the more general expression of the partial free enthalpy of component i when it finds itself in a body having mole numbers $n_1, n_2, \ldots n_N$.

Finally, in this section, let's find a way to express the total Gibbs free energy of a body in terms of the mole numbers and chemical potentials of the components, Equation (A2.10), on the next page. Now, the Gibbs free energy is an *extensive* property so if I know $G(T, p, n_1, n_2 \ldots n_N)$ for a body and then I assemble some number λ of these bodies together, then p and T do not change as these are intensive properties and the mole numbers are all multiplied by λ and the total Gibbs free energy is also multiplied by λ (because it is an extensive property):

$$G(T, p, \lambda n_1, \lambda n_2, \ldots \lambda n_N) = \lambda G(T, p, n_1, n_2 \ldots n_N)$$

(actually, λ does not need to be a whole number — if I extend the body by an amount λ, then the free energy is increased in the same ratio). We now invoke Euler's theorem for homogeneous functions of first order. Suppose a function f of N variables satisfies

$$f(\lambda x_1, \lambda x_2, \ldots \lambda x_N) = \lambda f(x_1, x_2, \ldots x_N)$$

We first exchange left- and right-hand sides and write $u_i = \lambda x_i$,

$$\lambda f(x_1, x_2, \ldots x_N) = f(u_1, u_2, \ldots u_N)$$

then differentiate each side with respect to λ, using the rule for differentiating a function of a function for the right-hand side,

$$f(x_1, x_2, \ldots x_N) = \sum_{i=1}^{N} \frac{\partial f}{\partial u_i} \frac{du_i}{d\lambda}$$

$$= \sum_{i=1}^{N} \frac{\partial f}{\partial u_i} x_i$$

$$= \sum_{i=1}^{N} \frac{1}{\lambda} \frac{\partial f}{\partial x_i} x_i$$

This is true for any λ, but if I choose $\lambda = 1$, then I get

$$f(x_1, x_2, \ldots x_N) = \sum_{i=1}^{N} \frac{\partial f}{\partial x_i} x_i$$

This is Euler's theorem which I now apply to the Gibbs function:

$$G(T, p, n_1, n_2 \ldots n_N) = \sum_{i=1}^{N} \left(\frac{\partial G}{\partial n_i} \right)_{T,p,n_{j \neq i}} n_i$$

and by comparison with (A2.7) I get

$$G = \sum_{i=1}^{N} \mu_i \, n_i \qquad (A2.10)$$

This states that the total free enthalpy of a phase having N components is equal to the sum of the number of moles of each component times its chemical potential.

From this, I can obtain the famous *Gibbs–Duhem equation*. If I take the total differential of (A2.10),

$$dG = \sum_{i=1}^{N} \mu_i \, dn_i + \sum_{i=1}^{N} n_i \, d\mu_i$$

and if I compare this with (A2.9), there results

$$-S dT + V dp - \sum_{i=1}^{N} n_i \, d\mu_i = 0 \qquad (A2.11a)$$

This is the Gibbs–Duhem equation which you can use to prove the *Gibbs phase rule* which is central to the interpretation of phase diagrams. At constant T and p, we then find

$$\sum_{i=1}^{N} n_i \, d\mu_i = 0 \qquad (A2.11b)$$

A.2.1   *Conditions for equilibrium in open systems*

Possibly the most significant statement in metallurgical and solution thermodynamics that you need to remember is this: *in equilibrium the chemical potential of each component is the same in all phases.* That is to say, if the body comprises more than one phase (and we will label the phases with subscripts α, β, ...) and the components are distributed among the phases so that the number of moles of component i in phase α is $n_{i\alpha}$, then as long as the phases are in contact and each component can diffuse throughout the body, then at equilibrium the chemical potential μ_i for component i is the same in each phase. This makes sense because it implies that if the chemical potential of component i is the same everywhere, then there's no gradient to drive diffusion and so all atoms stay where they are — equilibrium is reached. For example, a piece of Fe–C alloy at equilibrium within the $\alpha + \gamma$ field of the Fe–C phase diagram has a particular volume fraction of ferrite, having a very small concentration of carbon and a certain volume fraction of austenite with a large concentration of dissolved carbon. The concentrations of C in the two phases are very different because of the big difference in solubility of C in α-Fe and γ-Fe, but in equilibrium, $\mu_{C\alpha} = \mu_{C\gamma}$ — the chemical potential of carbon is the same everywhere.

For the curious, I will try and prove this for you. Suppose in a multiphase, multicomponent body I take an infinitesimal amount of component i, $dn_{i\alpha}$, from phase α and transfer it reversibly into phase β.[8] The amount of increase of component i in the β phase is,

---

[8]For example, I could have some Fe–C at the temperature at which austenite and ferrite are in equilibrium (at the carbon solubility phase boundary above A_1 in the Fe–C phase diagram). Then keeping T, p and the carbon concentrations fixed, I could take $dn_{Fe\,\alpha}$ moles of body centred cubic Fe (ferrite) and transfer them across the α / γ interface and rearrange them into the face centred cubic austenite. Alternatively I could reversibly move the α / γ interface in such a way that $dn_{Fe\,\alpha}$ moles of ferrite are rearranged into austenite. Actually, there is a complication in this thought experiment because once I have transformed some ferrite into austenite, I will have to also transfer an amount of carbon so that the austenite retains the carbon concentration appropriate to the tie line in the phase diagram; I gloss over this somewhat here, but as long as the amount I am transferring is *infinitesimal* in the sense of the differential calculus, then I can neglect the need to re-equilibrate the carbon in this process.

rather obviously, $dn_{i\beta} = dn_{i\alpha}$ while the change of mole number of component i in phase α is negative *viz.*, $-dn_{i\alpha}$. If I do this at constant temperature and pressure, then according to Equation (A2.9) the change in Gibbs free energy is

$$dG = -\mu_{i\alpha}dn_{i\alpha} + \mu_{i\beta}dn_{i\beta} = (\mu_{i\beta} - \mu_{i\alpha})dn_{i\alpha}$$

since (A2.9) must hold separately in each phase as each phase may be regarded as an open system and a part of the whole body. If the system was originally in equilibrium and the transfer is done reversibly, then

$$dG = 0 = (\mu_{i\beta} - \mu_{i\alpha})dn_{i\alpha}$$

and so, since $dn_{i\alpha} \neq 0$, it must be true that

$$\mu_{i\alpha} = \mu_{i\beta}$$

if the two phases are in equilibrium. We can extend this argument to any number of phases.

A.2.2 *Chemical potential in a closed system*

Paradoxically, we need to ask, what is the meaning of chemical potential in a closed system? This is because even if matter is not exchanged with the environment or a reservoir, the composition of a closed system may nonetheless change reversibly. A reversible chemical reaction may happen, for example, as a consequence of a change of temperature or pressure. In fact, Denbigh (see Further Reading in Appendix D) quotes Gibbs's definition of chemical potential: "If to any homogeneous mass we suppose an infinitesimal quantity of any substance to be added, the mass remaining homogeneous and its entropy and volume remaining unchanged, the increase of the energy of the mass divided by the quantity of substance added is the *potential* for the substance in the mass considered." This statement is exactly equivalent to the definition (A2.4). From a practical point of view, Gibbs's thought experiment is entirely possible if we imagine that although the additional matter will serve to increase the entropy of the mass, its total entropy can be brought back to what it was by the extraction of the appropriate amount of heat through contact with a heat bath. Similarly, the volume can be kept constant during the addition by the adjustment of the external pressure. A similar argument applies to (A2.7); in fact, the statement amounts almost trivially to the assertion that the free enthalpy is an extensive quantity — if I keep the *concentrations* of all components constant by simply extending the size of a closed system, then I will increase its free enthalpy. For example, if I add reagents in fixed proportion to the contents of a galvanic cell, I increase the amount of electrical work that it is able to do (see Section 9.3). In fact, a rather artificial device that we will need in Section 11.1 is to imagine a closed system as being simply a piece, of arbitrary size, taken by drawing a boundary within a larger whole. If I choose to extend that boundary, and hence increase the size of my system, I am changing the mole numbers of components but *not* by exchange of matter with external reservoirs.

Finally, let us consider again the formula

$$dU = TdS - pdV + \sum_{i=1}^{N} \mu_i dn_i \qquad (A2.5)$$

but now as applied to a *closed system*. The final terms then correspond to changes in the mole numbers, or concentrations, of the components due a chemical reaction (or even, presumably, a *nuclear* reaction). Of course for (A2.5) to apply to the closed system, before and after any such a change, the system must be initially in equilibrium, and remain in equilibrium (or quasi- or static equilibrium) throughout the process of change. In that case, since by definition the first term TdS is the change in heat content, the final term must be regarded as a "chemical" work term. This is exactly the work done by the passage of charge in a galvanic cell. Indeed, even at constant volume, it is clear that a system can perform work, lift a weight, in the absence of any change of volume as would be required in the operation, say, of a piston engine. The interpretation of the the last term in (A2.5) as work is only unambiguous in the case of a closed system (hence the significance of this sub-section); if the system is open and the change in mole numbers is as a consequence of exchange of matter with the environment or reservoir, then as matter is introduced or removed, it brings its own entropy and hence heat so that TdS is no longer the only heat term in (A2.5). The additional heat introduced by mass transport is sometimes called the "heat of transport" in diffusion problems.

A.3 Activity

Next, we need to know how the chemical potential relates to the concentration. Obviously, we design and process metal alloys and make up solutions of electrolyte using concentrations of the components but as we have seen what controls the thermodynamics and the kinetics at the fundamental level is the chemical potential. The best way to start this is simply to write down the answer. The chemical potential of component i in a particular phase (I'll leave out the subscript α unless it's needed) is

$$\mu_i = \mu_i^\circ + RT \ln a_i \qquad (A3.1)$$

μ_i° is the chemical potential in some "standard state" (mathematically, it's an integration constant) and depends only on temperature and pressure and a is the "activity". Where does this equation come from? At the most shallow level, it's simply a *definition* of activity, and still we need to find the connection between activity, a_i, and concentration, x_i, expressed as the mole fraction, n_i/n.[9] This is most generally expressed like this:

$$a_i = \gamma_i x_i$$

which may be a very complicated function since the so-called "activity coefficient", γ_i, is a function of temperature, pressure and the concentrations of all the components, not just component i:

$$\gamma_i = \gamma_i(T, p, x_1, x_2, \ldots x_N)$$

Thermodynamics can tell us nothing about how the activity depends on the concentrations. Ultimately, it is up to experiment or detailed

---

[9]When I use x_i for concentration, I mean the ratio of the number of moles or the number of atoms of component i to the total number of moles or the number of atoms in the mixture. If I write for the concentration c_i, then I mean the number of moles per unit volume. I hope it doesn't confuse you that I use the word concentration for both. Probably for x_i I should write "mole fraction," or "atom fraction." The relation between the two is

$$\frac{c_i}{x_i} = \frac{\rho(n_0 + \sum n_i)}{M_0 n_0 + \sum M_i n_i}$$

Here we imagine a number of solutes i dissolved in a solvent, labelled with the subscript 0; ρ is the density in $kg\,m^{-3}$ and M is the molar mass in $kg\,mole^{-1}$ (easily confused with relative molar mass — formerly known as molecular weight — which is one thousand times smaller).

204

Electrode and Corrosion Physics

atomic-scale theory to determine the relationship. For example, it follows from the Gibbs–Helmholtz equation that[10]

$$R\left(\frac{\partial \ln \gamma_i}{\partial(1/T)}\right)_{p,n_{j\neq i}} = h_i - h_i^\circ \qquad (A3.2)$$

where h_i° is the partial molar enthalpy in the same standard state as is applied to μ_i°. For an ideal solution, $\gamma =$ constant and so the left-hand side is zero, implying that the enthalpy of the component in solution is the same as in the pure substance at the same T and p, if the pure substance standard state is used. By integration of the Gibbs–Helmholtz equation, activity coefficients can be inferred from measured heats of solution.

The results of some electrochemical experiments are shown in Figure A.1. You can see that the activity does not at all show a straightforward relationship to the concentration, except to say that as the concentration increases, so does the activity. However, you see two very clear limits: for large concentrations, we see that

$$a_i = x_i \quad (x_i \to 1)$$

meaning that in the concentrated limit, the activity of the *solvent* is equal to its concentration, the activity coefficient being one, and

$$a_i = \gamma_i^\infty x_i \quad (x_i \to 0)$$

with γ_i^∞ constant, that is, in the infinitely dilute limit, the activity of the *solute* is proportional to its concentration. In this limit, the activity coefficient is a proportionality constant. These two limits are statements of, respectively, Raoult's law and Henry's law (first stated to refer to the partial pressures of gases).

---

[10]Combine (A1.5) with (A2.8),

$$G = H + T\left(\frac{\partial G}{\partial T}\right)_{p,n_i}$$

which can be recast as follows:

$$\left(\frac{\partial(G/T)}{\partial T}\right)_{p,n_i} = -\frac{H}{T^2}$$

which is the Gibbs–Helmholtz equation. Just divide G and H by the number of moles and insert (A3.1) with $a_i = \gamma_i x_i$ to get (A3.2).

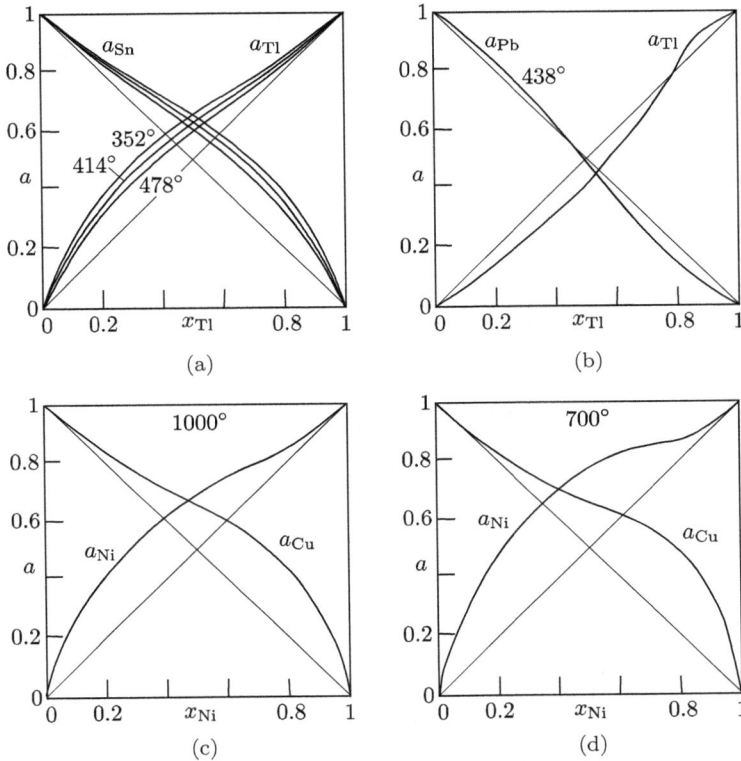

Fig. A.1.   Activity for some liquid and solid binary alloys as functions of mole fraction. When the concentration of a component tends to one, we then call this component the *solvent* for obvious reasons. (a) Liquid tin–thallium, (b) liquid lead–thallium, (c) solid copper–nickel, 1000°C (d) solid copper–nickel, 700°C. The straight lines show activities expected from Raoult's law, and significant deviations are observed, except that Raoult's law is always obeyed as the mole fraction tends to one. Generally, the deviation is positive, but note that both lead and thallium show negative deviations from Raoult's law in the dilute limits but positive deviations in the concentrated limits, that is, as they tend to become the solvent. In general, solutions tend towards the ideal as T increases.

Source: (a) and (b) are adapted with permission from Hildebrand, *J. Am. Chem. Soc.*, **51**, 462 (1929); (c) and (d) are adapted with permission from Rapp and Maak, *Acta Metall.*, **10**, 63 (1962).

The activities as a function of mole fraction in binary metal alloys shown in Figure A.1 are found by exploiting the Nernst equation in an electrochemical cell. A number of known alloy compositions are prepared and the e.m.f. is measured, which is proportional to the logarithm of the activity. Figures A.1(a) and A.1(b) show one of the earliest such measures on liquid thallium alloys. You note that tin–thallium shows a very strong "positive" deviation from Raoult's law, while lead–thallium shows a closer adherence to Raoult's law but having both positive and negative deviations. Later work, shown in Figures A.1(c) and A.1(d), was able to use a solid electrolyte and measure the activity in solid copper–nickel alloys. In all experiments just one activity is measured and the other is inferred from the Gibbs–Duhem equation (A2.11) which implies that

$$\sum_{i=1}^{N} x_i \, d \ln a_i = 0$$

so that the activity of one component can be inferred if the activity of the others is known.

In solid solutions with positive deviation, $a_i > x_i$, the solute behaves as if it were in a greater concentration than it actually is, and *vice versa*. Equation (A3.2) indicates that if $a_i > x_i$ (activity coefficient greater than one), then the mixing of the solute in the solvent is endothermic, meaning that the solute doesn't want to form a solid solution and the mixture has a tendency to phase separation or possibly spinodal decomposition. Conversely, if $\gamma_i < 1$, then the mixing is exothermic, very roughly meaning that the component i likes to form bonds with the atoms of the solvent; thermodynamics can say nothing about the origins of this which lie in quantum mechanics.

A.3.1 *Ideal mixture*

The "ideal mixture" or ideal solution, or ideal solid solution, like the ideal gas, is an idealisation but which fits the facts in many cases. They are characterised by a special form of the entropy of mixing, namely, if the two components are initially separated into two containers and subsequently allowed to mix at constant temperature, then the change in entropy is

$$\Delta S = R \sum_{i=1}^{N} n_i \ln \frac{n}{n_i} \qquad (A3.3)$$

$$= R \left(n \ln n - \sum_{i=1}^{N} n_i \ln n_i \right)$$

Here, n is the total number of moles and n_i is the number of moles of component i. For an ideal gas, we have from the ideal gas law (Dalton's law) that $n/n_i = p/p_i$, where p is the total pressure and p_i is the partial pressure of component i. So you can define the ideal solution as one whose entropy of mixing is taken by analogy with the ideal gas. You can see (A3.3) derived in a textbook on thermodynamics (look up "Gibbs's paradox" in Sommerfeld, see Further Reading in Appendix D).[11]

So when I take n_i moles of each of N components in the pure state at a given temperature and pressure, and I mix them to produce an *ideal* homogeneous single phase mixture of atoms, then the total free enthalpy of the body that I have created is the sum of the numbers of moles times the molar free enthalpies, $\mu_i^{\alpha,\bullet}$, of the pure substances in phase α plus the free enthalpy of mixing which is $-T\Delta S$ since the

---

[11] In the case of a binary mixture, you can derive (A3.3) from a statistical point of view. If the N atoms are arranged on the lattices of two perfect crystals, then the number of ways of arranging the atoms into two separated bodies is exactly **one**. When the atoms are randomly mixed on a common lattice in the alloy body, then the number of ways of arranging the N_a A-atoms and the N_b B-atoms is

$$W = \frac{(N_a + N_b)!}{N_a! N_b!} = \frac{N!}{N_a! N_b!}$$

So from the Boltzmann formula for the entropy, $S = k \ln W$, the entropy of mixing is

$$\Delta S = k \ln \frac{W}{\mathbf{one}} = k \left(\ln N! - \ln N_a! - \ln N_b! \right)$$

$$= -k \left(N_a \ln \frac{N_a}{N} - N_b \ln \frac{N_b}{N} \right)$$

using Stirling's slacker approximation $\ln x! \approx x \ln x - x$ for large x, and this is consistent with (A3.3).

solution is ideal. Hence,

$$G(T,p) = \sum_{i=1}^{N} n_i \left(\mu_i^{\alpha,\bullet}(T,p) - RT \ln \frac{n}{n_i} \right)$$

By comparison with Equation (A2.10), I see that the chemical potential of component i in the ideal mixture is

$$\mu_i = \mu_i^{\alpha,\bullet}(T,p) + RT \ln \frac{n_i}{n}$$

$$= \mu^{\bullet} + RT \ln \frac{n_i}{n} \qquad (A3.4)$$

so that $\mu^{\bullet}$ is the chemical potential of component i in its reference state: that is, in the pure substance in phase α (presumably but not necessarily the crystal structure adopted by the pure substance) at the temperature and pressure specified. So you can at least see where the structure of (A3.1) comes from: x_i is n_i/n and we replace x_i with a_i to recognise that not all mixtures are ideal in real life. In the case of the ideal gas, we have

$$\mathrm{d}G = V\mathrm{d}p - S\mathrm{d}T$$

and so at constant temperature,

$$\mathrm{d}G = V\mathrm{d}p = nRT\,\mathrm{d}\ln p; \quad \text{using the ideal gas law } pV = nRT$$

which on integration results in

$$\left(\frac{\partial G}{\partial n} \right)_{T,p} = \mu = \mu^{\ominus} + RT \ln \frac{p}{p^{\ominus}} \qquad (A3.5)$$

in which if the pressure is measured in bar ($p^{\ominus} = 10^5 \, \mathrm{N\,m^{-2}}$), then $\mu^{\ominus}$ is the chemical potential (free enthalpy per mole) in the standard state of one bar pressure and the temperature in question, and the pressure p is measured in units of one bar.

A.3.2 *Non-ideal mixture*

Up to now I have used the superscript ∘ applied to μ to denote some unspecified "standard state". In the case of the ideal gas, once we have integrated $\mathrm{d}G$, then the integration constant will depend on the boundary condition. If we choose to measure the pressure in bar, then (A3.5) indicates that $\mu^{\ominus}$ is the chemical potential of the ideal gas at one bar pressure and the temperature in question. We could have chosen different units of pressure, atmospheres or pascal, in which case the integration constant would be different. We are transferring weight from the constant term to the log term without altering the chemical potential in the right-hand side of (A3.5). This is called "choosing a standard state". Then, rather than the generic superscript $\circ$, we use the superscript $^{\ominus}$ to indicate that we are using the "one-bar standard state". The only restriction on the choice of standard state is that the standard chemical potential (the integration constant) should depend *only* on temperature and pressure.

In (A3.4), we used the superscript $^{\bullet}$ to denote the "pure substance standard state" so we rewrite (A3.1)

$$\mu_i = \mu_i^{\bullet} + RT \ln \gamma_i^{\bullet} x_i = \mu_i^{\bullet} + RT \ln a_i^{\bullet}$$

In the limit that the mixture tends to pure component i, $\gamma_i^{\bullet}$ tends to one,

$$\mu_i = \mu_i^{\bullet} + RT \ln \gamma_i^{\bullet} x_i \qquad \text{(A3.6a)}$$
$$\rightarrow \underbrace{\mu_i = \mu_i^{\bullet} + RT \ln x_i}_{\text{Raoult's law}} \; (x_i \rightarrow 1) \qquad \text{(A3.6b)}$$

At $x_i = 1$, the last term is zero ($\log 1 = 0$) and so $\mu_i = \mu_i^{\bullet}$ which identifies $\mu_i^{\bullet}$ as the chemical potential of pure substance i at the temperature and pressure of interest. From now on, we will refer to $a_i^{\bullet}$ as the *Raoultian activity* of component i, understanding that the Raoultian activity is referred to the pure solvent standard state. In the *infinitely dilute limit*, $x_i \rightarrow 0$, the activity coefficient becomes a constant, independent of concentration:[12]

$$\mu_i = \mu_i^{\bullet} + RT \ln \gamma_i^{\bullet} x_i \qquad \text{(A3.7a)}$$
$$\rightarrow \underbrace{\mu_i = \mu_i^{\bullet} + RT \ln \gamma_i^{\infty} x_i}_{\text{Henry's law}} \; (x_i \rightarrow 0) \qquad \text{(A3.7b)}$$

---

[12]The mathematically astute will notice that as x_i goes to zero so μ_i goes to minus infinity.

We would like, if possible, to be able to use concentration as a surrogate for the activity as then we don't have to worry about activity coefficients. If the concentration of a particular species is close to one, for example if it is the *solvent*, then we can invoke Raoult's law and use (A3.6b) as our chemical potential and $\mu^\bullet$ as our standard chemical potential. If on the other hand the species has a very small concentration, it may be that we can invoke Henry's law and assume a constant activity coefficient, γ^∞. We could then use (A3.7b) for the chemical potential. It is useful therefore to create a new, infinitely dilute, standard state in order to describe the activity of dilute solutes in a way that allows us to use concentration in place of activity in the dilute limit. To this end, we transfer the constant γ^∞ from the log term to the standard chemical potential so that Henry's law (A3.7b) becomes

$$\mu_i = \mu_i^\infty + RT \ln x_i \qquad (A3.7c)$$
$$= \mu_i^\infty + RT \ln h_i^\infty \qquad (A3.7d)$$

and

$$\mu_i^\infty = \mu_i^\bullet + RT \ln \gamma_i^\infty$$

is the chemical potential in the *Henrian* standard state and, since γ_i^∞ does not depend on composition, depends only on T and p. Of course the species may not be sufficiently dilute to obey Henry's law, but still it's worthwhile to refer its chemical potential to the Henrian standard state. Just as in acknowledging that if the solvent is not sufficiently concentrated, its chemical potential may be given by (A3.6a) in which the activity coefficient, $\gamma_i^\bullet$, depends on T, p and the concentrations of all species, so we write for the chemical potential of the *solute*

$$\mu_i = \mu_i^\infty + RT \ln \gamma^r x_i \qquad (A3.8a)$$
$$= \mu_i^\infty + RT \ln h_i^r \qquad (A3.8b)$$

in which h^r is the *rational Henrian activity* and γ^r is the activity coefficient which tends to one in the dilute limit just as $\gamma^\bullet$ tends to one in the concentrated limit of the Raoultian activity.[13]

---

[13]The "r" superscript stands for *rational*. I have been a bit unconventional and decorated my activities and activity coefficients with superscript symbols. This

These concepts are illustrated in Figure A.2. The Raoultian standard state (A3.7a) may be used over the whole range of concentration since the Raoultian activity, $a_i^{\bullet} = \gamma_i(T, p, x_1, x_2, \ldots x_N)x_i$, describes the whole solid curve and contains all of the non-ideality of the mixture. On the other hand, the Henrian standard state (A3.7c) is very convenient because we can use concentration directly as a surrogate for activity, but it is *only correct as long as the component is sufficiently dilute* for Henry's law to be valid; that is, the solid curve in Figure A.2 must coincide with the Henry's law straight line. Otherwise, we must multiply the atom fraction, x, by a Henrian activity coefficient γ^r, defined in (A3.8a), which like the Raoultian activity coefficient depends on T and p and on the concentration of all the components. Also the Henrian standard chemical potential, μ^{∞}, is impossible to visualise: it is the chemical potential for a solute at a mole fraction of one — at which point it has become pure solvent! On the other hand, the Raoultian standard chemical potential is clearly the molar free enthalpy of the pure substance, which may be conveniently be set to zero at standard temperature and pressure.

A.3.3 *Unimolal standard state*

The mole fraction, x_i, is the same as the atom fraction and is the number of moles (or atoms) of component i divided by the total number of moles (or atoms) in a substance. This is obviously a good measure of concentration in physical or metallurgical thermodynamics, but in electrochemistry and solution chemistry, it is usual to measure concentrations in moles per litre or moles per kilogram of solvent. By definition, one mole of a substance has a mass equal to its relative molar mass (molecular weight), M_W, in grams. Hence one kg of solvent contains $10^3/M_W$ moles. The mole fraction is

$$x_i = \frac{n_i}{n} \tag{A3.9}$$

and the concentration in moles per kg is

$$c_i = \frac{n_i}{M} \tag{A3.10}$$

---

is because in electrochemistry we want to use the *unimolal* standard state to be described in the next section and I therefore reserve un-superscripted symbols for that scale so that they can be raised to mathematical powers without notational clutter.

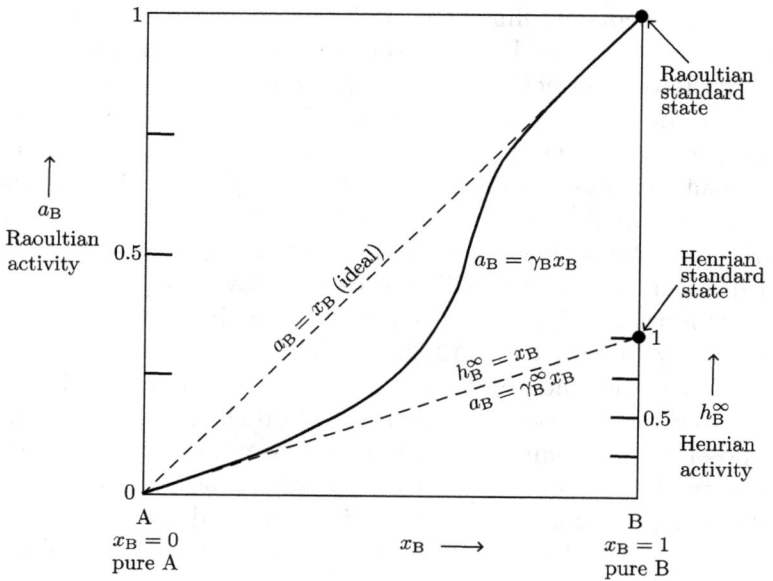

Fig. A.2.   Cartoon of the activity as a function of concentration (solid line) for a binary AB solution showing negative deviation from Raoult's law. Actual data are shown in Figure A.1. An ideal solution has $a = x$. The Raoultian activity, a, is equal to the concentration, x in the limit as $x \to 1$. The Henrian activity h^∞ is equal to x as $x \to 0$. Because the width of the abscissa is one, the slope of the Henrian activity as a function of x is γ^∞ and the straight line intersects the right-hand side ordinate at the Henrian standard state.

where M is the mass of the solvent in kg. If there are n moles of solvent, the mass of the solvent in kg is

$$M = 10^{-3} \times n\, M_W$$

and hence,

$$c_i = \frac{10^3}{M_W} \frac{n_i}{n} = x_i \frac{10^3}{M_W}$$

This is not dimensionless and so we finally define a quantity called the "molality"

$$m_i = \frac{c_i}{c^\circ} \approx x_i \frac{10^3}{c^\circ M_W} \tag{A3.11}$$

in which c° is the standard concentration of one mole per kilogram.

There is a small error which I have indicated in comparing x_i and c_i. In (A3.9), the denominator is the *total number of moles* $\sum_j n_j$ including the component i, whereas in (A3.10), the denominator is just the mass of the solvent. So (A3.10) is only correct in the dilute limit. All the same, we can still move weight from the log term to the standard chemical potential term, and introduce the *unimolal standard state*:[14]

$$\mu_i = \mu_i^{\square} + RT \ln h_i \tag{A3.12a}$$

$$= \mu_i^{\square} + RT \ln \gamma_i m_i \tag{A3.12b}$$

In this way, $\mu_i^{\square}$ is the standard unimolal chemical potential, h_i is the unimolal Henrian activity and γ_i is the unimolal Henrian activity coefficient. While μ^{∞} is truly a mythical state of a solute in unit mole fraction, $\mu^{\square}$ is the chemical potential of the solute in the concentration of one mole per kg of solvent at standard T and p — this is far from dilute.

For rather obvious reasons, the activity coefficient, γ^r, on the mole fraction scale is sometimes called *rational*, and the activity coefficient, γ, on the molality scale is sometimes called *practical*. The relations between the two scales are readily derived, and the standard chemical potentials are

$$\mu_i^{\infty} - \mu_i^{\square} = RT \ln \frac{10^3}{M_W} \frac{n}{n_0}$$

where n_0 is the number of moles of solvent. In the dilute limit,

$$\mu_i^{\infty} - \mu_i^{\square} = RT \ln \frac{10^3}{M_W}$$

The relation between practical and rational activity coefficients is

$$\frac{\gamma_i}{\gamma_i^r} = \frac{10^3}{M_W} \frac{x_i}{m_i}$$

In view of (A3.11), the two are equal in the dilute limit.

---

[14]I have been putting superscripts on h and γ up to now, in order to reserve the naked versions for the unimolal scale. This is because it is this scale that is mostly used in the main text, and in the following sections, where these are sometimes raised to some power so that a superscript would be clumsy. This is a bit unconventional.

A.3.4 *Chemical potential and activity of ionic species*

Some substances, acids, alkalis and salts, for example, *dissociate* into
their ions when dissolved in water. In the case of a dissolved salt,
such as sodium chloride, this will be completely ionised in water to its
constituent ions, Na^+ and Cl^-. We need to be careful of what can and
cannot be known of the individual ions — their chemical potentials
and activities are not *measurable*: only certain combinations are. We
consider a salt or acid which dissociates into anions and cations,

$$C_{z_+}A_{z_-} \rightleftharpoons z_+C^{z+} + z_-A^{z-}$$

in which the cation, C, has a charge number z_+, and the anion, A, has
a charge number z_-. We imagine that m moles of the substance are
dissolved in n_0 moles of solvent. If not all the substance is dissociated,
then let m_u be the number of moles of undissociated substance. The
numbers of moles of the ions in solutions will then be

$$m_+ = z_+ (m - m_u) \qquad (A3.13a)$$

$$m_- = z_- (m - m_u) \qquad (A3.13b)$$

The total differential (A2.9) in this case will be

$$dG = -SdT + Vdp + \mu_u dm_u + \mu_+ dm_+ + \mu_- dm_- + \mu_0 dn_0$$

According to (A2.7), the chemical potential of the ions will be

$$\mu_+ = \left(\frac{\partial G}{\partial m_+}\right)_{T,p,m_-,n_0}$$

$$\mu_- = \left(\frac{\partial G}{\partial m_-}\right)_{T,p,m_+,n_0}$$

the subscript 0 referring to the solvent (water). This change in mole
numbers cannot be realised in practice because even an infinitesimal
increase in the number of positive ions while keeping the number
of negative ions fixed will give rise to a large electric field that will
swamp the change in energy. We now need to look into what can

be known. In terms of (A3.13) at constant T and p,

$$dG = \mu_u dm_u + z_+\mu_+ (dm - dm_u) + z_-\mu_- (dm - dm_u) + \mu_0 dn_0$$
$$= (\mu_u - z_+\mu_+ - z_-\mu_-) dm_u + (z_+\mu_+ + z_-\mu_-) dm + \mu_0 dn_0$$
$$\text{(A3.14)}$$

Now all the differentials refer to quantities that *can* be varied independently without charging up the solution. *Firstly*, consider the case where n_0 and m are fixed. Then at equilibrium,

$$\left(\frac{\partial G}{\partial m_u} \right)_{T,p,m,n_0} = 0$$

from which according to (A3.14),

$$\mu_u = z_+\mu_+ + z_-\mu_-$$

which expresses the condition after m moles of the substance have been added to n_0 moles of the solvent and the dissociation has proceeded to equilibrium. *Secondly*, now relax the constraint on m and n_0 but allow that as these are varied this is done slowly enough that equilibrium is maintained in the process so the degree of dissociation is always as in the previous condition. Then the first term in (A3.14) vanishes by virtue of the last equation and

$$dG = (z_+\mu_+ + z_-\mu_-) dm + \mu_0 n_0$$

Now it is possible to define the chemical potential of the substance through the variation of m at fixed n_0 while maintaining the equilibrium dissociation of the substance:

$$\mu = \left(\frac{\partial G}{\partial m} \right)_{T,p,n_0}$$

By comparison of the last two equations, it is evident that

$$\mu = z_+\mu_+ + z_-\mu_- \qquad \text{(A3.15)}$$

The conclusion is that individual chemical potentials of ions cannot be defined but *a combination weighted by charge numbers* amounts to the chemical potential of the dissolved substance.

As to the activity coefficients of individual ions, these could be defined via the following equations:

$$\mu_+ = \mu_+^\circ + RT \ln \gamma_+ m_+ \qquad \text{(A3.16a)}$$

$$\mu_- = \mu_-^\circ + RT \ln \gamma_- m_- \qquad \text{(A3.16b)}$$

Putting (A3.16) into (A3.15), we get

$$\mu = z_+ \mu_+^\circ + z_- \mu_-^\circ + RT \ln \gamma_+^{z+} \gamma_-^{z-} m_+^{z+} m_-^{z-}$$

Since the chemical potentials on the left-hand sides of (A3.16) are not measurable, neither are the quantities on the right-hand sides. However, since μ is measurable, then so is the linear combination

$$\mu^\circ = z_+ \mu_+^\circ + z_- \mu_-^\circ \qquad \text{(A3.17)}$$

and so is the product $\gamma_+^{z+} \gamma_-^{z-}$. We let $z = z_+ + z_-$ and we define the *mean ion activity coefficient*, $\gamma_\pm$, through the formula

$$\mu = z_+ \mu_+^\circ + z_- \mu_-^\circ + RT \ln \gamma_\pm^z m_+^{z+} m_-^{z-} \qquad \text{(A3.18)}$$

In this way, we have identified

$$\gamma_\pm^z = \gamma_+^{z+} \gamma_-^{z-} \qquad \text{(A3.19)}$$

as a proper measurable combination of ion activity coefficients. Even though the individual ion molalities, m_+ and m_-, are known from the concentration of our electrolyte, we can go further and define a *geometric mean ion molality*:

$$m_\pm = \left(m_+^{z+} m_-^{z-} \right)^{1/z} \qquad \text{(A3.20)}$$

and finally (A3.18) takes on the simple form

$$\mu = z_+ \mu_+^\circ + z_- \mu_-^\circ + zRT \ln \gamma_\pm m_\pm \qquad \text{(A3.21a)}$$

$$= \mu^\circ + zRT \ln \gamma_\pm m_\pm \qquad \text{(A3.21b)}$$

A.3.5   *Electrolyte chemical potential and activity*

Electrochemistry is fraught with difficulty because the student would like to use equilibrium thermodynamics and is hampered by the problem that there are no well-defined thermodynamic properties of charged species, or substances which carry a net charge, so that the charge neutrality demanded by thermodynamic theory cannot be guaranteed. What is most needed is a measure of the activity of an ion. We achieve a "working value" of this in (A3.26) at the end of this section. Suppose that a substance is totally dissociated as is, for example, NaCl or hydrochloric acid, HCl, or an alkali such as NaOH, then (A3.13) simplifies to $m_+ = z_+ m$ and $m_- = z_- m$ and these are known from the amount of salt or acid we've thrown into the water. Suppose we are dealing with an electrolyte such as sodium chloride, then as in (A3.11) we can work out the molality of the electrolyte, m_{el}. If the solution is not ideal, then the activity in the *practical Henrian scale* is

$$h_{el} = \gamma_{el} m_{el}$$

and γ_{el} is the activity coefficient of the electrolyte in that scale. It is surely uncontroversial to write for the chemical potential of the electrolyte in solution using the unimolal practical standard,

$$\mu_{el} = \mu_{el}^{\square} + RT \ln h_{el} \qquad (A3.22a)$$

$$= \mu_{el}^{\square} + RT \ln \gamma_{el} m_{el} \qquad (A3.22b)$$

We accept from the previous section which is thermodynamically rigorous that *at least in principle* there exist single ion chemical potentials and activities. And, as in (A3.16), that we can also write for the individual ions,

$$\mu_+ = \mu_+^{\square} + RT \ln h_+$$

$$\mu_- = \mu_-^{\square} + RT \ln h_-$$

which defines "thermodynamic" ion activities:

$$h_+ = \gamma_+ z_+ m_{el} = \gamma_+ m_+ \qquad (A3.23a)$$

$$h_- = \gamma_- z_- m_{el} = \gamma_- m_- \qquad (A3.23b)$$

So if we now write (A3.17) for the electrolyte, we'll have

$$\mu_{el} = z_+\mu_+ + z_-\mu_-$$

$$\mu_{el}^\circ = z_+\mu_+^\circ + z_-\mu_-^\circ$$

then in terms of activities, after subtracting,

$$\mu_{el} - \mu_{el}^\circ = z_+\left(\mu_+ - \mu_+^\circ\right) + z_-\left(\mu_- - \mu_-^\circ\right)$$

$$= z_+RT\ln h_+ + z_-RT\ln h_-$$

$$= RT\ln h_{el}$$

whence,

$$h_{el} = h_+^{z_+}h_-^{z_-} \tag{A3.24a}$$

$$\equiv h_\pm^z \tag{A3.24b}$$

where $z = z_+ + z_-$. This means that we can express the activity of the electrolyte in terms of a *geometric mean ion activity*, $h_\pm$. As we saw in the previous section, we can define a mean ion activity coefficient (A3.19), $\gamma_\pm^z = \gamma_+^{z_+}\gamma_-^{z_-}$, and using (A3.23), it follows by simple algebra that

$$h_\pm = \gamma_\pm z_+^{z_+} z_-^{z_-} m_{el}^z \tag{A3.25}$$

We also understand that (A3.22b) and (A3.12b) are expressions for the same chemical potential. It then follows that as long as the electrolyte is fully dissociated,

$$\gamma_{el} = \gamma_\pm^z \, z_+^{z_+} z_-^{z_-} \, m_{el}^{z-1}$$

In this way, the relation between the activity coefficient of the electrolyte and the mean ion activity coefficient varies with the concentration (molality). For a one-to-one electrolyte, such as NaCl or HCl, at a molality of one (the standard state),

$$\gamma_\pm = \sqrt{\gamma_{el}}$$

So far so good, and thermodynamically rigorous. However, the activity of individual ions cannot be determined experimentally. It is sometimes necessary to assume some values of these, in which case *leaving thermodynamics aside* momentarily a modification of (A3.23) is possible in order to define "working values of the ion activities":

$$h_+ = \gamma_\pm z_+ m_{el} = \gamma_\pm m_+ \tag{A3.26a}$$

$$h_- = \gamma_\pm z_- m_{el} = \gamma_\pm m_- \tag{A3.26b}$$

A.3.6 *Measurement of mean ion activity coefficient*

We consider again the dissociation of an electrolyte into its cations and anions in solution in water:

$$C_{z_+}A_{z_-} \rightleftharpoons z_+C^{z+} + z_-A^{z-}$$

The mean ion activity coefficient can be measured in an electrochemical cell, as long as one electrode is reversible with respect to the anion and one with respect to the cation.[15]
   The cell diagram

$$\mathrm{Pt, H_2 \ (g)|HCl \ (molality} = m)|Hg_2Cl_2|Hg}$$

represents a hydrogen electrode connected to a "calomel electrode". (Hg_2Cl_2 is the substance calomel which was used widely as a medicine for a huge number of diseases, until it was realised in the late 19th century that mercury was poisoning the patients.) The cell reaction is

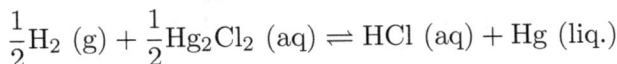

$$\frac{1}{2}H_2 \ (g) + \frac{1}{2}Hg_2Cl_2 \ (aq) \rightleftharpoons HCl \ (aq) + Hg \ (liq.)$$

According to Equation (9.3.6) in the main text,

$$\mu_{HCl} + \mu_{Hg} - \frac{1}{2}\mu_{H_2} - \frac{1}{2}\mu_{Hg_2Cl_2} = -F\mathcal{E}$$

Using (A3.21), we have

$$\mu_{HCl} = \mu_{H^+}^{\ominus} + \mu_{Cl^-}^{\ominus} + 2RT\ln\gamma_{\pm}m_{\pm}$$

and according to (A3.20), in this instance,

$$m_{\pm} = m$$

the molality of the electrolyte, HCl. Assuming that the hydrogen is an ideal gas,

$$\frac{1}{2}\mu_{H_2} = \frac{1}{2}\mu_{H_2}^{\ominus} + \frac{1}{2}RT\ln p_{H_2}$$

---

[15]See Section 2.1 in the main text. *Reversible* means that the electrode reaction is unique and rapid. For example, in the lemon lamp (Chapter 1 in the main text), the Zn electrode is not reversible because the electrode reaction may either be $Zn \rightarrow Zn^{++} + e^-$ or $2H^+ + e^- \rightarrow H_2$, which can short circuit the cell so that the "resistance" of the interphase is not infinite (see Figure 2.3 in the main text).

Table A.1.   Mean ion activity coefficient of aqueous hydrochloric acid at increasing concentration.

m_{HCl}	0.001	0.005	0.01	0.05	0.1	0.5	1	2
x_{HCl}	3.6×10^{-5}	1.8×10^{-4}	3.6×10^{-4}	0.0018	0.0036	0.018	0.036	0.072
$\gamma_\pm$	0.966	0.928	0.905	0.830	0.796	0.757	0.809	1.009

where p_{H_2} is the pressure in bar. We now have

$$RT \ln \frac{\gamma_\pm^2 m^2}{\sqrt{p_{H_2}}} - F\mathcal{E}^\circ = -F\mathcal{E}$$

where the *standard e.m.f.* of the cell is defined by

$$-F\mathcal{E}^\circ = \mu_{H^+}^\square + \mu_{Cl^-}^\square + \mu_{Hg}^\bullet - \frac{1}{2}\mu_{H_2}^\ominus - \frac{1}{2}\mu_{Hg_2Cl_2}^\bullet$$

which is the standard free enthalpy change of the cell reaction. If we maintain the pressure of hydrogen at one bar, then

$$2RT \ln m + F\mathcal{E} = F\mathcal{E}^\circ - 2RT \ln \gamma_\pm$$

and the two terms in the left-hand side are *measurable*. The constant $F\mathcal{E}^\circ$ can be determined by varying the molality (concentration) of the acid and recording the open-circuit voltage (e.m.f.) of the cell. This can be extrapolated to zero molality at which $\gamma_\pm = 1$ (although there are subtleties needed to avoid the infinity in the logarithm — see Denbigh, in Further Reading, Appendix D) and it is found that $\mathcal{E}^\circ$ at 25°C is 0.26796 V. Once that is known, then at any concentration, the equation can be used to find $\gamma_\pm$ of the electrolyte.

Table A.1 shows measurements for HCl which show how deviation from Henry's law is observed as the concentration increases.

In this table, (A3.11) is used to compare the practical molality with the rational mole fraction concentrations. Note that the activity coefficient is less than one in the dilute limit (of course tending to one in the infinitely dilute limit), that it becomes initially smaller as the concentration increases, but that at large concentrations it becomes greater than one. Note also that one molal HCl has a mean molal activity of about 0.8. This is why the concentration of HCl in the standard hydrogen electrode is used at 1.2 molal to achieve "unit activity of the hydrogen ion," see Footnote 5 of Section 9.4.

Appendix B

List of Symbols

Symbol	Description	Where defined
ϵ_0	Permittivity of free space, $1/4\pi\epsilon_0 = 9 \times 10^9$ (N m^2 C^{-2})	
ϵ	Permittivity	
e	Elementary charge, $+1.602 \times 10^{-19}$ coulomb	
h	Planck constant	
k	Boltzmann constant, also $\beta = 1/kT$	
L	Avogadro constant	
M_W	Relative molar mass (molecular weight)	(A3.11)
R	Gas constant = Boltzmann constant × Avogadro constant	$R = 8.3$ J/mole K
F	Faraday constant = Elementary charge × Avogadro constant	$F = 96500$ C/mol
K	Equilibrium constant	(9.1.3)
$n(\mathbf{r})$	Number of electrons per unit volume	Section 3.1
n	Total number of moles, $\sum_i n_i$	
n	Number of electrons involved in an electrochemical reaction	(9.3.6)

(*Continued*)

(*Continued*)

Symbol	Description	Where defined
p	Dipole moment per unit area	(3.2.4) *et seq.*
ϕ, ϕ_m, ϕ_s	Inner, or Galvani (electric) potential (in metal or solution)*	(3.2.1) *et seq.*
ψ, ψ_m, ψ_s	Outer, or Volta (electric) potential (in metal or solution)	(5.1)
χ, χ_m, χ_s	Dipole (electric) potential (in metal or solution)*	$\phi = \psi + \chi$ (5.1)
$^A\Delta^B X$	$X_A - X_B$ for a potential X	(3.2.3) and Footnote 2
$\mathcal{E}$	Electromotance, e.m.f., $\mathcal{E} = {}^{m'}\Delta^m\phi$	(9.3.5) and above
$\mathcal{E}^\circ$	Standard e.m.f. of an electrochemical cell	(9.4.2) and (9.6.3)
$\Delta_c\phi$	Compensation potential (e.m.f.) of cell with air gap	(10.1.1) and Section 3.3
q_m, q_s	Charge on metal, electrolyte	Chapter 5
$\overline{V}$	Average Hartree plus external potential (energy) relative to DFT zero	Figure 3.1
μ	Electron chemical potential relative to DFT zero	(3.1.1)
μ_e, $\mu_e^{Pt}\cdots$	Fermi level* ($\mu_e = \mu - \overline{V}$)	(3.1.2)
$\tilde{\mu}_e$, $\tilde{\mu}_e^{Pt}\cdots$	Electron electrochemical potential, $\tilde{\mu}_e = \mu_e - e\phi_m$*	(3.4.4)
W^∞	Electron work function of a metal, relative to the potential at infinity;* $W^\infty = -\tilde{\mu}_e$ for an uncharged metal	(3.4.3a)

(Continued)

Symbol	Description	Where defined
W	Electron work function of a metal relative to the vacuum potential (at the Bockris point)	(3.4.3b)
W_s	Electron work function of a solution	(7.1)
$W_i^{s,\circ}$	Ion work function of species i in solution in standard state	(8.2.1)
P_B	Bockris point: a place "just outside the surface of a phase"	Figure 5.2 and Chapter 6
U	Potential energy	(13.6.2) and Section 13.7
E_{tot}	Total energy (in DFT)	Section 3.1
U	Internal energy	(A1.1)
F	Free energy (Helmholtz function)	(A1.3)
G	Free enthalpy (Gibbs function)	(A1.5)
$\Delta_f G^\circ$	Standard free enthalpy of formation	(9.1.4)
ΔG°	Standard free enthalpy change in a chemical reaction	(9.1.2)
K	Equilibrium constant	(9.1.3)
$\Delta_{sol}^\infty G_i^\circ$	Standard free enthalpy of solvation of species i^*	(8.2.5)
$\Delta_{sol}^r G_i^\circ$	Standard real free enthalpy of solvation of species i	(8.2.7)
$\mu^\bullet$	Free enthalpy per mole of pure substance	(A3.4)
μ^∞	Infinitely dilute standard chemical potential	(A3.7c)

(Continued)

(*Continued*)

Symbol	Description	Where defined
$\mu^{\square}$	Unimolal standard chemical potential	(A3.12) and Section 9.2
$\mu^{\ominus}$	Standard chemical potential of gas at one bar pressure	(A3.5)
$\tilde{\mu}_i^{g,\circ}$	Electrochemical potential of charged species i in standard gas phase*	(8.2.1)
$\tilde{\mu}_i^{s,\circ}$	Electrochemical potential of charged species i in standard solution*	(8.2.3)
$\alpha_i^{s,\circ}$	Standard real potential of species i in solution	(8.2.1)
μ_i	Chemical potential of species i	(A2.4) and (A2.7)
μ_+, μ_-	Ion chemical potential*	After (A3.22)
$\tilde{\mu}_i$	Electrochemical potential,* $\tilde{\mu}_i = \mu_i + z_i F \phi_s$	(4.1.1)
α	Real potential, $\alpha_i = \mu_i + z_i F \chi_s$	(4.2.1)
α,β	Symmetry factors, $\alpha = 1 - \beta$	(14.4.3)
z_i	Charge number of species i	(4.1.1)
z_+, z_-	Charge number of cation and anion	Above (A3.13)
a	Activity, generally $a_i = \gamma_i(T, p, x_1, x_2, \ldots x_N) \times x_i$	(A3.1) and Figure A.2
$a_i^{\bullet}$	Raoultian activity of species i, $a_i^{\bullet} = \gamma_i^{\bullet} x_i$	(A3.6) and above
x_i	Concentration of species i as mole or atom fraction	(A3.9)
c_i	Concentration of species i in moles per kg solvent	(A3.11)

(*Continued*)

Symbol	Description	Where defined
m_i	*Molality:* normalised concentration of species i (moles per kg solvent)	(A3.11)
$m_\pm$	Geometric mean ion molality	(A3.20)
h^∞	Infinitely dilute Henrian activity $h^\infty = x$	(A3.7d)
h^r	Rational Henrian activity (referred to mole fraction, x)	(A3.8b)
h	Unimolal, practical Henrian activity (referred to molality, m)	(A3.12a)
h	Partial molar enthalpy (also v, f, g)	Section A.2
h_+, h_-	"Thermodynamic" practical activity of an ion*	(A3.23)
h_+, h_-	"Working" practical activity of an ion	(A3.26)
$h_\pm$	Geometric mean ion activity	(A3.24)
h_{el}	Practical Henrian activity of electrolyte	(A3.22)
$\gamma^\bullet$	Raoultian activity coefficient (not constant, but $\gamma^\bullet \to 1$ as $x \to 1$)	(A3.7a)
γ^∞	Henry's law constant, $a = \gamma^\infty x$ for $x \to 0$	After (A3.7c)
γ^r	Rational Henrian activity coefficient (not constant, but $\gamma^r \to \gamma^\infty$ as $x \to 0$)	(A3.8a)
γ	Practical Henrian activity coefficient (not constant)	(A3.12b)

(*Continued*)

(*Continued*)

Symbol	Description	Where defined
γ_+, γ_-	Single ion activity coefficient*	(A3.16)
γ_{el}	Practical Henrian activity coefficient of electrolyte	(A3.22)
$\gamma_\pm$	Mean ion activity coefficient	(A3.19)
γ	Surface tension, interfacial excess free enthalpy	(11.1.2)
Γ_i	Interfacial excess of component i	(11.1.7a)
u, v, s, n	Layer extensive quantities per unit area	Before (11.1.3a)
V	Voltage, volume	
σ	Surface charge density	(3.3.1) and (11.2.4)
ρ	Volume charge density, electrical resistivity	Section 12.1 and Chapter 15
C	Electrical capacitance	(3.3.1)
$\mathcal{C}$	(*Differential*) *capacity*, capacitance per unit area, $\mathcal{C} = d\sigma/dV$	(11.2.5)
$K^{\ddagger}$	Equilibrium constant of a reactant and its activated complex	Section 13.6
$\Delta G^{\ddagger}$	Free enthalpy of activation	(13.1.1)
$\Delta H^{\ddagger}$	Enthalpy of activation	After (13.1.1)
$\Delta S^{\ddagger}$	Entropy of activation	After (13.1.1)
$\Delta G^{\circ\ddagger}$	Standard free enthalpy of activation	(13.6.3a)
$\Delta H^{\circ\ddagger}$	Standard enthalpy of activation	(13.6.3b)
$\Delta S^{\circ\ddagger}$	Standard entropy of activation	(13.6.3b)
$\Delta G_e^{\ddagger}$	Electronation free enthalpy of activation	Section 14.2

(*Continued*)

Symbol	Description	Where defined
$\Delta G_d^{\ddagger}$	De-electronation free enthalpy of activation	Section 14.3
$\Delta\phi_{eq}$	Equilibrium interphase potential difference	Section 9.2 and (14.1.2)
$\Delta\phi_{eq}^{\circ}$	Standard equilibrium interphase potential difference	(14.7.2)
k_1	First order rate coefficient	(13.1.1)
k_e	Electronation rate coefficient	(14.7.1)
k_d	De-electronation rate coefficient	(14.7.1)
$\bar{c}$	Areal concentration (moles per unit area)	Section 14.2
$\bar{a}_{kink}$	Activity of a metal surface kink site (mole m^{-2})	Section 14.3
r	Rate of reaction	(13.3.3)
r_e^0	Zero-field rate of electronation (mole m^{-2} s^{-1})	(14.2.1)
r_d^0	Zero-field rate of de-electronation (mole m^{-2} s^{-1})	(14.3.1)
r_e	Rate of electronation (mole m^{-2} s^{-1})	(14.3.1)
r_d	Rate of de-electronation (mole m^{-2} s^{-1})	(14.2.1)
ν_e	Electronation frequency prefactor (s^{-1})	Section 14.2
ν_d	De-electronation frequency prefactor (s^{-1})	Section 14.3
i_e	Electronation current density (amp m^{-2})	Footnote 4 of Section 14.5

(*Continued*)

(*Continued*)

Symbol	Description	Where defined
i_d	De-electronation current density (amp m^{-2})	Footnote 4 of Section 14.5
i_e^{eq}	Equilibrium electronation current density (amp m^{-2})	(14.2.2)
i_d^{eq}	Equilibrium de-electronation current density (amp m^{-2})	(14.3.2)
i_0	$= i_e^{eq} = i_d^{eq}$: equilibrium exchange current density	(14.5.2)
i	Butler–Volmer current density (amp m^{-2})	(15.1)
$\mathcal{E}_{cor}$	Corrosion potential (V)	Figure 16.4
i_{cor}	Corrosion current density (amp m^{-2})	Figure 16.4

Note: *Denotes not measurable.

Some Worked Problems

Problem 1.1

Which of the following chemical reactions is oxidation, reduction or neither?

(1) $Fe^{++} + e^- \rightarrow Fe$
(2) $Fe^{+++} + 3(OH)^- \rightarrow Fe(OH)_3$
(3) $O_2 + 4H^+ + 4e^- \rightarrow 2H_2O$
(4) $\frac{1}{2}H_2 \rightarrow H^+ + e^-$
(5) $Mg + H_2O \rightarrow MgO + 2H^+ + 2e^-$

Solution

(1) Reduction
(2) Neither
(3) Reduction
(4) Oxidation
(5) Oxidation (obviously!)

Problem 1.2

Construct a lemon lamp and use a simple multimeter in place of the bulb. Record the voltage and note the polarity.

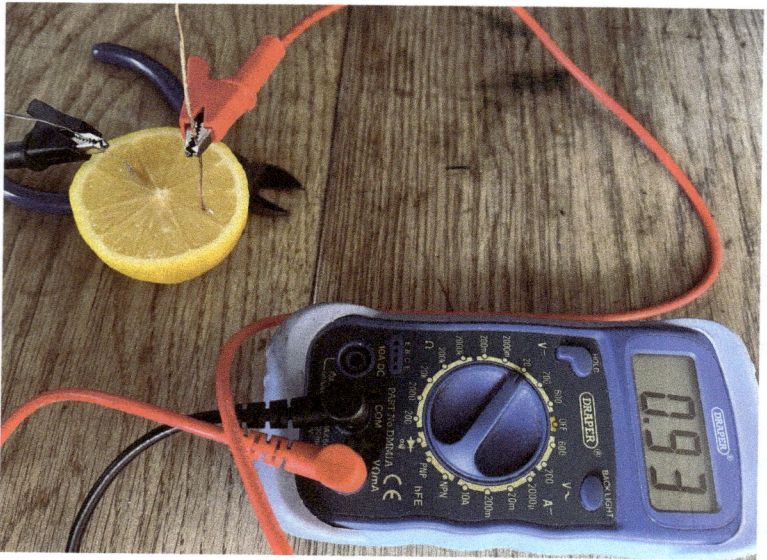

Fig. C.1.   My lemon lamp.

Solution

Figure C.1 is a photo of my lemon lamp. You see it is recording 0.93 V. The copper wire is positive and the zinc strip is negative.

Problem 2.1

Another standard electrode which is reversible within some range of potentials is the saturated calomel electrode (SCE). A platinum wire dips into liquid mercury which is in contact with a paste of calomel, Hg_2Cl_2, which in turn is in contact with a saturated aqueous solution of potassium chloride. Write out the cell diagram for the electrochemical cell of Figure 2.4 if the standard hydrogen electrode (SHE) is replaced by an SCE. Use Pt rather than Cu wire. Use a double vertical bar to indicate the salt bridge.

Solution

$$Pt \mid Zn \mid ZnSO_4 \parallel KCl \mid Hg_2Cl_2 \mid Hg \mid Pt'$$

Problem 2.2

(i) Suppose the left-hand cell in Figure 2.4 is replaced with a reversible standard calomel electrode. Will current flow if the electrodes are electrically connected? (ii) Suppose the cell is made from two ideally polarisable electrode half cells. Will current flow? (iii) How do you account for the current in the lemon lamp?

Solution

(i) Yes, each half cell acts as a resistor with no capacitance. (ii) No, each half cell acts as a capacitor and will simply charge up without current flowing. (iii) The previous answers are true *in principle*, but half cells are not ideally polarisable or ideally reversible and so current does leak out through the resistor element in Figure 2.6.

Problem 3.1

We see that from Equation (3.1.1) the chemical potential, which is a central quantity in all that follows, is identified as a mere Lagrange multiplier constraining the total number of electrons. Does this worry you or strike you as cheating? There are other examples in which a central quantity in a theory turns out to be originally an innocuous seeming Lagrange multiplier. Can you think of any?

Solution

In Boltzmann's statistical mechanics, he maximises the entropy, or equivalently the number of "complexions", W, over microcells in P. and T. Ehrenfest's "μ-space", subject to constraints on the number of particles, N, and the total energy, U. He finds

$$\ln W_{\max} = N \ln N + \alpha N + \beta U$$

where α and β are Lagrange multipliers which ensure the constancy of the number of particles and the total energy. It turns out then that

$$\alpha = \ln \frac{z}{N}$$

and

$$\beta = \frac{1}{kT}$$

where z is the partition function, T is the absolute temperature, and k is the Boltzmann constant. Thus, two of the most central quantities in statistical thermodynamics started out life as humble Lagrange multipliers.

Another example comes from the dynamics of an incompressible fluid. There is a real difficulty in imagining how pressure can exist or be defined in a strictly incompressible continuum. The resolution rests on a formulation of the equations of motion in fluid dynamics within Hamilton's mechanics after which the pressure emerges as precisely the Lagrange multiplier needed to constrain the fluid to be incompressible.

Problem 3.2

Define a functional derivative mathematically. For an electronic charge density, $-en(\mathbf{r})$ as a function of position, $\mathbf{r}$, where $-e$ is the charge on the electron, the Hartree energy is[1]

$$E[n] = \frac{1}{2}e^2 \iint \mathrm{d}\mathbf{r}_1 \mathrm{d}\mathbf{r}_2 \frac{n(\mathbf{r}_1)n(\mathbf{r}_2)}{|\mathbf{r}_1 - \mathbf{r}_2|}$$

which is the classical electrostatic self-energy of the charge distribution. The notation $E[n]$ indicates that E is a functional of $n(\mathbf{r})$. Find the first functional derivative of E with respect to n and hence an expression for the Hartree potential energy.

Solution

A function $f(x)$ takes a number x and turns it into another number. A functional $F[f(x)]$ takes a function $f(x)$ and turns it into a number (or into another function).

---

[1] In atomic units, $4\pi\epsilon_0 = 1$.

Think of the number $F[n(\mathbf{r})]$ as a function of many variables, $f(n(\mathbf{r}_1)n(\mathbf{r}_2)\cdots n(\mathbf{r}_m))$. Then a small change in F is

$$\delta F = F\left[n(\mathbf{r}) + \delta n(\mathbf{r})\right] - F\left[n(\mathbf{r})\right]$$

$$= \sum_{i=1}^{m} \frac{\partial f}{\partial n(\mathbf{r}_i)}\delta n(\mathbf{r}_i)$$

and taking the limit as $m \to \infty$,

$$\delta F = \int \xi(\mathbf{r})\delta n(\mathbf{r})d\mathbf{r}$$

This defines the functional derivative, in the sense of the fundamental theorem of calculus,

$$\xi(\mathbf{r}) \doteq \frac{\delta F}{\delta n}(\mathbf{r}) \qquad\qquad\qquad (S3.2.1)$$

To get the functional derivative of the Hartree energy,

$$E[n] = \frac{1}{2}e^2 \int\int \frac{n(\mathbf{r}_1)n(\mathbf{r}_2)}{|\mathbf{r}_1 - \mathbf{r}_2|}d\mathbf{r}_1 d\mathbf{r}_2$$

$$\delta E[n] = E[n + \delta n] - E[n]$$

$$= \frac{1}{2}e^2 \int\int \frac{[n(\mathbf{r}_1) + \delta n(\mathbf{r}_1)][n(\mathbf{r}_2) + \delta n(\mathbf{r}_2)]}{|\mathbf{r}_1 - \mathbf{r}_2|}d\mathbf{r}_1 d\mathbf{r}_2$$

$$- \frac{1}{2}e^2 \int\int \frac{n(\mathbf{r}_1)n(\mathbf{r}_2)}{|\mathbf{r}_1 - \mathbf{r}_2|}d\mathbf{r}_1 d\mathbf{r}_2$$

$$= e^2 \int\int \frac{n(\mathbf{r}_1)\delta n(\mathbf{r}_2)}{|\mathbf{r}_1 - \mathbf{r}_2|}d\mathbf{r}_1 d\mathbf{r}_2 + \mathcal{O}(\delta n)^2$$

Comparing with Equation (S3.2.1), we get

$$\frac{\delta E}{\delta n}(\mathbf{r}_2) = e^2 \int \frac{n(\mathbf{r}_1)}{|\mathbf{r}_1 - \mathbf{r}_2|}d\mathbf{r}_1$$

$$\doteq V_{\mathrm{H}}, \quad \text{the Hartree potential energy}$$

Problem 3.3

In density functional theory, we solve a problem of interest specified by the number, N, of electrons and an external potential, v_{ext}. Use first-order perturbation theory of the many electron wave function, Ψ, to show that for a small variation in the number of electrons at fixed external potential,

$$\left(\frac{\partial E}{\partial N}\right)_{v_{\text{ext}}} = \mu$$

which is the chemical potential described in Section 3.1, if E is the total energy.

Solution

We make a change in the external potential,

$$\Delta V_{\text{ext}} = \sum_i \Delta v_{\text{ext}}(\mathbf{r}_i)$$

In other words, each of the electrons i experiences a small change $\Delta v_{\text{ext}}(\mathbf{r}_i)$. We don't need to consider only the ground state in this development. If the solutions to the Schrödinger equation without the perturbation (superscript 0) are labelled k,

$$\hat{H}\Psi_k^0 = E_k^0\Psi_k^0$$

then the first-order change in the energy E_k when the change in v_{ext} is applied is

$$\Delta E_k = \langle \Psi_k^0 | \Delta V_{\text{ext}} | \Psi_k^0 \rangle$$

$$= \int n_k(\mathbf{r}_1)\Delta v_{\text{ext}}(\mathbf{r}_1)\mathrm{d}\mathbf{r}_1 \qquad \text{(S3.3.1)}$$

In the ground state, $k = 0$, this is the Hellmann–Feynman theorem. To get the second line from the first, we observe that

$$\sum_{i=1}^{N} \int \cdots \int \mathrm{d}\mathbf{r}_1 \ldots \mathrm{d}\mathbf{r}_N \overline{\Psi}_k^0 \Psi_k^0 v_{\text{ext}}(\mathbf{r}_i)$$

$$= N \int \cdots \int \mathrm{d}\mathbf{r}_1 \ldots \mathrm{d}\mathbf{r}_N \overline{\Psi}_k^0 \Psi_k^0 v_{\text{ext}}(\mathbf{r}_1)$$

that is, N identical terms, because all the $\mathbf{r}_i$ are dummy variables of integration and can be interchanged freely because $\overline{\Psi}_k^0 \Psi_k^0$ doesn't change when you interchange variables. Then, we use the definition

$$n_k(\mathbf{r}_1) \doteq N \int \cdots \int ds_1 ds_2 d\mathbf{r}_2 \ldots ds_n d\mathbf{r}_N \left|\Psi_k^0\right|^2$$

asserting that the probability that there is an electron in $d\mathbf{r}_1$ is $n(\mathbf{r}_1)d\mathbf{r}_1$, having marginalised, or integrated out, the spin degrees of freedom and the coordinates of the remaining $N - 1$ electrons.

The expression for the first-order change, (S3.3.1), is exact for both ground and excited states of the many body wavefunction and independent of the choice of exchange and correlation functional — even the exact one, if it could be found. For the ground state density, we write $n_k = n$.

Let us now concentrate on the ground state — the only state accessible in density functional theory. We have just seen that for a change ΔV_{ext} in the external potential, the first-order change in the total energy is

$$\delta E = \langle \Psi_0^0 | \Delta V_{\text{ext}} | \Psi_0^0 \rangle = \int n(\mathbf{r}) \Delta v_{\text{ext}}(\mathbf{r}) d\mathbf{r}$$

From our solution to Problem 3.2 on functional derivatives, we then have

$$\left(\frac{\delta E}{\delta v_{\text{ext}}}\right)_n = \left(\frac{\delta E}{\delta v_{\text{ext}}}\right)_N = n(\mathbf{r}) \tag{S3.3.2}$$

If the density is held constant, then so is the number of electrons. We know that $E = E[N, v_{\text{ext}}(\mathbf{r})]$, so

$$dE = \left(\frac{\partial E}{\partial N}\right)_{v_{\text{ext}}} dN + \int \left(\frac{\delta E}{\delta v_{\text{ext}}(\mathbf{r})}\right)_N dv_{\text{ext}}(\mathbf{r}) d\mathbf{r} \tag{S3.3.3}$$

Then using (see Section 3.1),

$$E[n] = F[n] + \int n(\mathbf{r}) v_{\text{ext}}(\mathbf{r}) d\mathbf{r}$$

we see that, since $F[n]$ does not depend explicitly on the external potential,

$$dE = \int \left(\frac{\delta E}{\delta n(\mathbf{r})}\right)_{v_{\text{ext}}} dn(\mathbf{r}) d\mathbf{r} + \int \left(\frac{\delta E}{\delta v_{\text{ext}}(\mathbf{r})}\right)_n dv_{\text{ext}}(\mathbf{r}) d\mathbf{r} \tag{S3.3.4}$$

We also have the Euler–Lagrange equation

$$\left(\frac{\delta E}{\delta n}\right)_{v_{ext}} = \mu$$

and

$$dN = \int dn(\mathbf{r})d\mathbf{r}$$

Putting these last two into Equation (S3.3.4) and using (S3.3.2), we get

$$dE = \mu\,dN + \int n(\mathbf{r})dv_{ext}(\mathbf{r})d\mathbf{r}$$

and comparing this with (S3.3.3), we obtain

$$\mu = \left(\frac{\partial E}{\partial N}\right)_{v_{ext}}$$

which is the *chemical potential*. Sometimes $-\mu$ is called the *electronegativity* since out of equilibrium electrons will be attracted to regions of low chemical potential.

It may worry you that I cannot add an infinitesimal amount of electron charge to a system since electrons are quantised in units of $-e$. Indeed, the total energy of an atom, say, as a function of number of electrons is not a smooth function but piecewise linear with slopes being the electron affinity or ionisation potential, which are not equal. The resolution of this in density functional theory is to consider a grand canonical ensemble of systems. To see how this is done, consult N. D. Lang and W. Kohn, *Phys. Rev. B*, **3**, no. 4, 1215 (1971) and J. F. Janak, *ibid.*, **18**, no. 12, 7165 (1978).

Problem 3.4

Show that Equation (3.4.3b) is consistent with Equation (3.5.1).

Solution

Equation (3.4.3b) is

$$W = -\mu_e + e\chi \tag{3.4.3b}$$

and

$$\tilde{\mu}_e = \mu_e - e\phi = \mu_e - e\psi - e\chi$$

with $\phi = \psi + \chi$. Then,

$$\mu_e = \tilde{\mu}_e + e\psi + e\chi$$

and putting this into (3.4.3b), we get (3.5.1):

$$W = -\mu_e + e\chi = -\tilde{\mu}_e - e\psi \qquad (3.5.1)$$

You will see both these expressions for the work function in the literature.

Problem 3.5

Consider this thought experiment. You prepare a single crystal of tungsten having both (100) and (110) faces. You remove an electron from the Fermi level through a (100) face and take it to at least 10^{-6} cm from the surface. You then move it to opposite a (110) face and return it through there back to the Fermi level so that the initial and final states have the same total energy. On the other hand, you have gained 0.62 eV of energy, that being the difference in work functions belonging to the two crystal surfaces. That energy might have performed useful work and you have constructed a perpetual motion machine. Resolve this conflict.

Tungsten crystals are prepared and two specimens are polished having (100) and (110) faces respectively with surface areas of A cm^2. These are placed parallel to each other with an air gap of amount x cm separating them, and a piezoelectric device is used to move them at a constant velocity $v = dx/dt$. The crystals are wired to an ammeter so that as they move the current, I, flowing between the two specimens is measured. Find a formula for I in terms of the work function difference between the two faces. At some moment during the experiment, a current of one microamp is measured while the piezoelectric device records a speed of 0.1 ms^{-1}. The area of each of the two crystal faces is 1 cm^2. At that moment, what is the separation of the two faces?

Solution

The only resolution to this paradox is that there must be electric fields present that we need to do work against so that the total work done vanishes. These fields can only be explained by surface charges in addition to surface dipoles that are created by the perturbation of the atomic and electronic structures of the crystal caused by the creation of the surfaces. We must take it that since less than the expected amount of work is needed to withdraw the electron from the lower work function face, that face must carry a net negative charge density. The crystal as a whole must be neutral, so to compensate the (110) face must be positively charged. It is also possible that the edges and corners are charged, but then there must be some explanation for why some are positive and others negative, whereas by symmetry, they are expected to be identical.

According to Section 3.2, the contact electric potential difference is

$$\Delta\psi = \frac{\sigma x}{\epsilon_0}$$

The total charge, q, is σA, so

$$q = \frac{A\epsilon_0\Delta\psi}{x}$$

and the current is

$$I = \frac{dq}{dt} = -A\epsilon_0\Delta\psi\frac{v}{x^2} = A\epsilon_0\frac{\Delta W}{e}\frac{v}{x^2}$$

using the relation between contact potential and work function of Section 3.5. This is the formula we are looking for. Working in SI units,

$$A = 10^{-4}\,\mathrm{m^2}, \quad \epsilon_0 = 8.854\times10^{-12}\,\mathrm{C^2\,N^{-1}m^{-2}},$$

$$e = 1.602\times10^{-19}\,\mathrm{C}$$

$$\Delta W = 0.62\times1.602\times10^{-19}\,\mathrm{J}, \quad I = 10^{-6}\,\mathrm{A}, \quad v = 0.1\,\mathrm{ms^{-1}}$$

you find a distance of 7.4 microns.

To read further on the Kelvin probe, see *Electrochimica Acta*, **53**, 290 (2007).

Problem 3.6

The "jellium" model of a simple metal smears out the positive charge of the combined nuclei and core electrons into a uniform density, n_0, against which background moves an electron gas of equal and opposite charge density. Unlike the Sommerfeld free electron gas, in jellium the electron–electron interaction is taken beyond the Hartree picture, to higher levels including local density approximation and beyond. When jellium is cleaved to produce a surface, the positive charge drops discontinuously to zero at the surface, while the electron gas spills out to some extent into the vacuum, as discussed in Chapter 3. This situation is illustrated in Figure C.2, which shows the positive charge density as constant up to the surface and the negative electron density spilling into the vacuum to an extent described by a screening length, λ. Note that the electron gas in the metal screens out the exposed positive charge at the surface in a decaying "Friedel oscillation."

Figure C.2 is appropriate to the average electron density of sodium. By the crude approximation of replacing the electron density

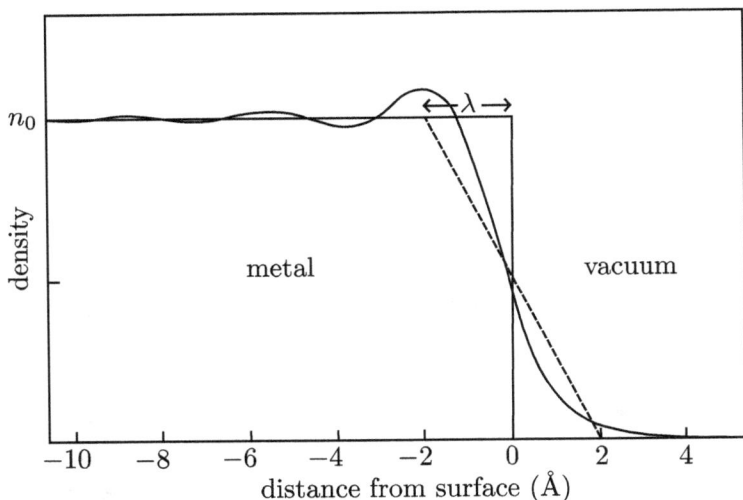

Fig. C.2. Electron density and density of positive background at the jellium surface.
Source: Adapted with permission from N. D. Lang and W. Kohn, *Phys. Rev. B*, **3**, 1215 (1971).

with the dashed straight line, calculate the surface dipole potential. This may seem over crude, but approximating with a simple exponential decay gives the same answer (see Schmickler and Santos, Chapter 4, Problem 3). Make some comments on your findings in relation to the known dipole potential of water and to the cartoons of Chapter 3.

Solution

Take the x-direction to be normal to the surface. The straight line segments when added to the background are described by

$$\rho(x) = en_0 + (-e)\frac{1}{2}n_0\left(1 - \frac{x}{\lambda}\right) = \frac{1}{2}en_0\left(1 + \frac{x}{\lambda}\right)$$

for $x < 0$, and

$$\rho(x) = (-e)\frac{1}{2}n_0\left(1 - \frac{x}{\lambda}\right) = -\frac{1}{2}en_0\left(1 - \frac{x}{\lambda}\right)$$

for $x > 0$. As shown in Chapter 3, we need to integrate $x\rho(x)/\epsilon_0$ between $-\lambda$ and λ. The result is

$$\chi = \frac{1}{2}\frac{en_0\lambda^2}{\epsilon_0}$$

I work out the valence electron density in sodium from the crystal lattice constant, a_{bcc}, as

$$n_0 = \frac{2}{a_{\text{bcc}}^3} = 2.5 \times 10^{28}\text{ m}^{-3}$$

as there are two atoms per body centred cubic unit cell and one valence electron per atom. If I take $\lambda = 2\text{Å}$ and use the fundamental constants, I get

$$\chi = 0.9\text{ V}$$

Note that this is nearly 50 times larger than the dipole potential at the surface of water, see Problem 10.1. However, this is smaller than the work function of sodium which is about 3.6 V. This implies by comparison with Equation (3.4.3b) that μ_e must be *negative* and this is in contrast to the way I have drawn cartoons, such as Figure 3.4. Clearly, I have drawn them wrongly, at least for sodium. However, the cartoons allow the situation to be illustrated more clearly, but the fact is that the Fermi energy may well fall *below* the average electrostatic energy, $\overline{V}$. The reason for this is that in the Hartree approximation, or Sommerfeld theory, electrons are regarded as non-interacting and so their average electrostatic potential energy is much larger than actual, since electron correlation and exchange ensure that in their complex dynamics electrons stay well away from each other. We assert that the electron digs itself an "exchange and correlation hole" which exposes the positive background around the electron and it can be shown that the electron plus its hole are in total an electrically *neutral* "quasi-particle." Therefore, in spite of the huge Coulomb repulsion between electrons, they actually behave in real life like weakly interacting particles and so their eigenvalues may be more negative than $\overline{V}$.

Problem 7.1

Go ahead and calculate the electron work function of an electrolyte. Table C.1 contains some data (in eV) taken from *J. Phys. Chem. A*, **122** 7464 (2018) and wikipedia.

Note that free enthalpies of solvation are large and negative. This highlights the excellent property of water as a polar solvent, able very effectively to screen a large charge by the orientation of the electric dipoles of the water molecule. Of course, as you see, the reduction in electrostatic energy is greater for the higher charged ion. Compare your result with the ionisation energy of liquid water, which is about 11.6 eV.

Table C.1.   Real solvation free enthalpies of doubly and triply charged ions in water.

Ion	$\Delta_{sol}^{r}G(M^{++})$	$\Delta_{sol}^{r}G(M^{+++})$	I_3
Fe	−20.0	−45.4	30.6
Cr	−20.1	−47.0	31.0
Ag	−20.2	−48.5	34.8
Co	−20.8	−47.9	33.5
Mn	−19.1	−46.8	33.7
Cu	−21.6	−51.6	36.8
Ni	−21.4	−49.9	35.2

Note: All data are in eV.

Table C.2.   Results of the calculation in Problem 7.1.

Redox couple	W_s	$\mathcal{E}^\circ$	$\mathcal{E}^\circ + \mathcal{E}_{SHE}^{abs}$
Fe^{++}/Fe^{+++}	5.2	0.78	5.2
Cr^{++}/Cr^{+++}	4.0	−0.41	4.0
Ag^{++}/Ag^{+++}	6.5		
Co^{++}/Co^{+++}	6.4	1.9	6.4
Mn^{++}/Mn^{+++}	6.0	1.5	5.9
Cu^{++}/Cu^{+++}	6.8		
Ni^{++}/Ni^{+++}	6.7		

Notes: For each redox couple in the first column, the second column is the ion work function calculated from data in table C.1; the third column shows the couple's std. e.m.f., so you can confirm that for any two couples the ion work function difference is the same as the difference between their std e.m.f.; the fourth column shows the std e.m.f. of the couple added to that of the SHE, confirming that this is the same as the the ion work function of column 2.

Solution

You find 5.2 eV (Fe), 4.0 eV (Cr), 6.5 eV (Ag), 6.4 eV (Co), 6.0 eV (Mn), 6.8 eV (Cu) and 6.7 eV (Ni). See Table C.2. Schmickler and Santos point out that once the electron work function with respect to a particular redox couple is known, then the work function with respect to any other redox couple is found by simply adding the difference in standard potentials of the two couples. This is born out in Table C.2 for the few cases of the standard potential that I can find. Schmickler and Santos further point out that given the absolute e.m.f. of the standard hydrogen electrode, one can in fact find the electron work function for any redox couple, simply by adding its standard e.m.f. to that of the SHE, namely 4.44 V (see Section 10.2, Equation (10.2.5)). This is demonstrated in the fourth column in Table C.2. You could fill in the blanks in column 3 by this method, demonstrating that you can deduce the standard e.m.f. of the redox couples that I couldn't find from the measured free enthalpies and ionisation potentials in Table C.1.

The work function is roughly about 2 eV larger than that in a metal. It is interesting that W_s is about half the ionisation energy of liquid water, measured to be 11.6 eV (*J. Phys. Chem. Lett.*, **11** 1789 (2020)). Maybe this is because the "carrier" is screening the charge of the otherwise bare electron and hence lowering the amount of work needed to remove it from the water.

Problem 9.1

Professor R. M. Lynden-Bell told me of a useful way to clean tarnished silver. I put the silver items into a plastic bowl with some screwed up aluminium foil, having them in good physical contact. I put a slack handful of washing soda in the bowl and cover everything in hot water. Gas is evolved, and after some time, the foil appears to become distressed, thin and brittle and the silver comes out with a bright surface and the tarnish removed. Can you explain at least qualitatively what is going on?

Solution

The simple explanation is that, as you see in the electrochemical series, aluminium is more base than silver. The soda acts as electrolyte and the electrical contact between the silver and the foil allows an electrochemical cell to be established. The aluminium becomes the anode and corrodes, while at the silver, the cathode, the tarnish is reduced to metal. Silver tarnish is usually a film a few tens of nanometres thick of copper chloride and silver sulphide. The sulphur may come from the polluted atmosphere or from the yolk of an egg. The chloride comes from contact with salt. Pure silver will tarnish much less readily than stirling silver which is alloyed with copper. The tarnished appearance is a consequence of light interference patterns from variations in the thickness of the film. This is just a qualitative answer. For a proper chemical description of the cleaning process, see J. Novakovic, *et al.*, *Int. J. Electrochem. Sci.*, **8**, 7223 (2013).

Problem 10.1

Confirm that indeed the surface dipole potential of water is 20 mV. A recent estimate of $\Delta_{sol}^{\infty} G_{H+}^{\circ}$ is -11.37 eV using high precision quantum chemistry (G. J. Tawa, *et al.*, *J. Chem. Phys.*, **109**, 4852 (1998)). On the other hand, from mass spectroscopy, it is measured to be -11.45 eV (M. D. Tissandier, *et al.*, *J. Phys. Chem. A*, **102**, 7787

(1998)). Use the latter result and then comment on the outcome if you use the first.

Solution

In Section 10.1, we found

$$W_{H+}^{s,o} = \mu_{H+}^{g,o} - \alpha_{H+}^{s,o} = \frac{1}{2}4.52 + 13.61 - 4.44 = 11.43 \text{ eV}$$

Equation (8.2.5) is, in absolute microscopic units,

$$\Delta_{sol}^{\infty}G_{H+}^{o} = -\left(W_{H+}^{s,o} + e\chi_s\right)$$

and so

$$\chi_s = \frac{1}{e}(11.45 - 11.43) = 20 \text{ mV}$$

If I use the theoretical quantum chemistry calculation, I get $\chi_s = -60$ mV. We are subtracting large numbers here to get a small one and so even the sign is in doubt. On the other hand, very authoritative authors agree that the surface dipole potential is either 170 mV or 140 mV. The first is from Jun Cheng and Michiel Sprik, *Phys. Chem. Chem. Phys*, **14**, 11245 (2012), who use 11.36 eV for the proton work function (taken from Trassati) and -11.53 eV for the proton solvation free enthalpy, but this appears to be a misprint as the citation they use is the same as given earlier, namely -11.45 eV, in which case they actually mean to get 90 mV. The estimate of 140 mV is from W. Ronald Fawcett, *Langmuir*, **24**, 9868 (2008), but he uses 4.21 eV for the dissociation energy of the hydrogen molecule, which is curious since everyone seems to agree it's 4.52 eV. Interestingly, Fawcett mentions that quantum chemical calculations of the surface dipole potential of water indeed find it to be negative but also of a much larger magnitude than any of the earlier estimates. It looks as if the last word has not yet been spoken on this.

Problem 11.1

Ethanol is added to pure water to a mole fraction concentration of 0.01 at 25°C; this reduces the surface tension of water from 72 J m^{-2}

to 60 J m^{-2}. At the surface of the water, what is the excess of ethanol with respect to solvent?

Solution

We can approximate this as a two-component single-phase problem by neglecting the concentrations of water and ethanol in the air in contact with the water. Then, we use Equation (11.1.8) in the simple form:

$$d\gamma = -\Gamma \, d\mu$$

where μ is the chemical potential of the solute, ethanol. Then, in the dilute limit, we can assume

$$\mu = \mu_0 + RT \ln x$$

where $x = 0.01$ is the mole fraction of ethanol. From this, we have

$$d\mu = RT \, d\ln x = RT \frac{dx}{x}$$

and hence,

$$\Gamma = -\frac{1}{RT} x \frac{d\gamma}{dx}$$
$$\approx -\frac{1}{RT} x \frac{\Delta\gamma}{\Delta x}$$
$$= -\frac{1}{8.314 \times 298} 0.01 \frac{60 - 72}{0.01 - 0}$$
$$= 0.005 \text{ mol m}^2$$

Problem 11.2

In the case of a single phase, we defined an excess of component i with respect to the volume,

$$\Gamma_i^{(V)} = [n_i/V] = \left(n_i - \frac{V}{V^\alpha} n_i^\alpha\right)$$

and I claimed that this is independent of the volume of the system at a given interfacial area, A. Show that extending the length of the rectangle in Figure 11.1 by an amount ΔL by adding homogeneous phase, α, indeed does not alter the value of $\Gamma_i^{(V)}$. Do this using the properties of determinants.

Solution

I write the excess of component i as a determinant as in (11.1.10):

$$\Gamma_i^{(V)} = \frac{1}{V^\alpha} \begin{vmatrix} \mathsf{n}_i & \mathsf{v} \\ n_i^\alpha & V^\alpha \end{vmatrix} \tag{S11.2.1}$$

If I extend the length, say to the left in Figure 11.1, by ΔL, then I add a volume,

$$A\Delta L$$

where A is the area of the interface, or cross-sectional area of the system, Figure 11.1(b). Then I have that the number of moles of component i becomes

$$n_i^\alpha \to n_i^\alpha + \frac{n_i^\alpha}{V^\alpha} A\Delta L$$

that is, it is increased by the number of moles per unit volume of homogeneous phase, α, times the increase in volume. Dividing through by A we get

$$\mathsf{n}_i \to \mathsf{n}_i + \frac{\Delta L}{V^\alpha} n_i^\alpha$$

The layer volume becomes

$$\mathsf{v} \to \mathsf{v} + \frac{1}{A} A\Delta L = \mathsf{v} + \frac{\Delta L}{V^\alpha} V^\alpha$$

So by adding a length of homogeneous phase α, I have modified the determinant in (S11.2.1), so it has become

$$\begin{vmatrix} \mathsf{n}_i & \mathsf{v} \\ n_i^\alpha & V^\alpha \end{vmatrix} + \frac{\Delta L}{V^\alpha} \begin{vmatrix} n_i^\alpha & V^\alpha \\ n_i^\alpha & V^\alpha \end{vmatrix}$$

but the second determinant is zero, so the result is unchanged.

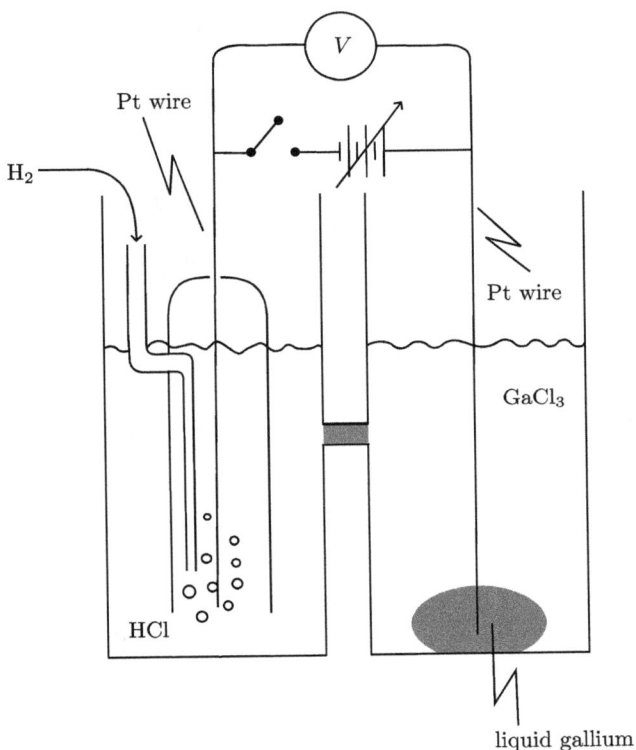

Fig. C.3. Standard hydrogen electrode connected via salt bridge, high impedance voltmeter and variable power source in parallel to a cell of liquid gallium electrode in gallium chloride.

Problem 11.3

As represented in Fig. C.3, I imagine a standard hydrogen electrode connected through a voltmeter and a salt bridge to a half cell containing a solution of gallium chloride and a drop of molten Ga at a temperature of $35°C$. A platinum wire dips into the liquid Ga and is connected to the voltmeter as indicated. The SHE is kept at $25°C$.

(1) If the molal concentration of the $GaCl_3$ is 10^{-4}, find the voltage recorded at the voltmeter. What is the polarity?
(2) If I close the switch, I can use the variable voltage supply and voltmeter to assert a Galvani potential difference, ϕ, between the platinum wires. By assuming that the SHE is ideally reversible

and that the electrolytes in either half cell are mutually in equilibrium (ignore the temperature difference), show that if the applied potential is V, then the change in surface energy of the gallium drop is

$$d\gamma = -\sigma\,dV - \Gamma_{Ga^{3+}}\,d\mu_s$$

where σ is the excess metal surface charge density, $\Gamma_{Ga^{3+}}$ is the interfacial excess of Ga ions in the electrolyte and μ_s is the chemical potential of the electrolyte. Note that all terms on the right-hand side are measurable.

(3) As a positive potential is applied to the right-hand electrode, describe what change, if any, is observed in the shape of the liquid gallium drop.

Solution

(1) Consulting the electrochemical series, we find that on SHE scale $\mathcal{E}^0 = -0.53$ V. Then from the Nernst equation,

$$\mathcal{E} = \mathcal{E}^0 + \frac{RT}{3F}\ln h_{Ga^{3+}}$$

That is,

$$\mathcal{E} = -0.53 + \frac{8.314 \times 303}{3 \times 96485}\ln 10^{-4} = -0.61 \text{ V}$$

The right-hand side is positive.

(2) We will focus on "derivation number two" of the Lippmann equation in the text. As we change the Galvani potential, ϕ, in the gallium, the interfacial free enthalpy changes according to

$$d\gamma = -\sigma\,d\phi - \sum_i \Gamma_i\,d\mu_i = -\sigma\,d\phi - \Gamma_{Ga^{3+}}\,d\tilde{\mu}_{Ga^{3+}} - \Gamma_{Cl^-}\,d\tilde{\mu}_{Cl^-}$$

The crux of the argument is that the chloride ions are in equilibrium across the two half cells by virtue of the salt bridge and so we may claim that the action of the ideally reversible SHE is to maintain the electrochemical potential of the chloride ions constant. Hence,

$$d\tilde{\mu}_{Cl^-} = 0$$

The chemical potential of the electrolyte in the right-hand cell is (see Section 8.1)

$$\mu_s = \tilde{\mu}_{Ga^{3+}} + 3\tilde{\mu}_{Cl^-}$$

Therefore,

$$d\tilde{\mu}_{Ga^{3+}} = d\mu_s$$

In view of this, we arrive at the result

$$d\gamma = -\sigma\,dV - \Gamma_{Ga^{3+}}\,d\mu_s$$

where the applied change in voltage is the same as the change in the Galvani potential in the gallium, assuming that the SHE is ideally non-polarisable.

(3) If the potential is made more positive, this will reduce the surface energy of the gallium which will cause the drop to flatten out since its shape reflects the competition between gravity flattening it and its surface energy tending to make it spherical.

Problem 13.1

One mole of particles occupies a cubic box whose sides are 0.3 m in length at 25°C. What is the average spacing between the particles? Compare this with the thermal de Broglie wavelength,

$$\Lambda = \frac{h}{\sqrt{2\pi m k T}}$$

if the particles are electrons or helium atoms. Comment on the result in terms of the ratio:

$$\frac{\text{volume per particle}}{\Lambda^3} = \frac{V}{N\Lambda^3}$$

The only degree of freedom of a monatomic ideal gas is translation. Using the result that

$$U = \frac{3}{2}NkT$$

from equipartition (although this result can be obtained from the partition function), show using $F = -kT \ln Z$ that the entropy of N atoms of an ideal gas occupying a volume, V, is

$$S = Nk \ln \left(e^{5/2} \frac{V}{N\Lambda^3} \right)$$

Does this go to zero as the temperature goes to zero? If not, why not? You will use

$$Z = \frac{z^N}{N!}$$

Why do I include the factor $1/N!$?

Solution

The volume per particle is V/N, where $N = L$ is the Avogadro constant. Then, the mean distance between particles is

$$\left(\frac{V}{N} \right)^{1/3} = \left(\frac{0.027}{6.022 \times 10^{23}} \right)^{1/3} = 3.55 \times 10^{-9} \text{ m}$$

Putting in numbers,

$$k = 1.381 \times 10^{-23} \text{ J K}^{-1}, \quad T = 298 \text{ K}, \quad h = 6.6261 \times 10^{-34} \text{ J s}$$
$$m_{He} = 6.65 \times 10^{-27} \text{ kg}, \quad m_e = 9.1094 \times 10^{-31} \text{ kg}$$

you find that the thermal de Broglie wavelengths are 4.32×10^{-9} m for the electron and 5.05×10^{-11} m for the helium atom. You see that He at one bar pressure may be safely treated as a gas of classical particles, not so the electron gas since the thermal de Broglie wavelength is of the same order as the particle spacing so their de Broglie wavepackets are overlapping. In other words,

$$\frac{V}{N\Lambda^3} \sim \mathcal{O}(1)$$

We begin with the partition function,

$$Z = \frac{1}{N!} z^N = \frac{1}{N!} V^N \left(\frac{2\pi mkT}{h^2}\right)^{\frac{3}{2}N}$$

and find the free energy,

$$F = -kT \ln Z$$

$$= -kT \left\{ -\ln N! + N \ln V + N \ln \left(\frac{2\pi mkT}{h^2}\right)^{\frac{3}{2}N} \right\}$$

$$= -NkT \left\{ -\ln N + 1 + \ln V + \ln \left(\frac{2\pi mkT}{h^2}\right)^{\frac{3}{2}N} \right\}$$

where I have used the slacker Stirling approximation, $\ln N! = N \ln N - N$. Now I separate out the fundamental constants from the variables,

$$F = -NkT \left\{ 1 + \ln \left(\frac{V m^{3/2} T^{3/2}}{N}\right) + \ln \left(\frac{2\pi k}{h^3}\right)^{3/2} \right\}$$

I know that the internal energy is

$$U = \frac{3}{2} NkT$$

and $F = U - TS$, or,

$$S = \frac{1}{T}(U - F)$$

Therefore,

$$S = Nk \left\{ \frac{5}{2} + \ln \left(\frac{V m^{3/2} T^{3/2}}{N} \right) + \ln \left(\frac{2\pi k}{h^3} \right)^{3/2} \right\}$$

$$= Nk \ln \left(\frac{e^{5/2} V m^{3/2} T^{3/2} (2\pi k)^{3/2}}{N h^3} \right)$$

$$= Nk \ln \left(e^{5/2} \frac{V}{N} \frac{(2\pi m k T)^{3/2}}{h^3} \right)$$

$$= Nk \ln \left(e^{5/2} \frac{V}{N \Lambda^3} \right)$$

These are some of the various forms in which you will find the Sackur–Tetrode formula written down. There are at least two answers to why the entropy does not go to zero with the temperature as it is supposed to. The simple answer is that there is no ideal gas at low temperature; the second and more rigorous is that the summation has not been done quite right. You can follow the details in Sommerfeld (see Further Reading in Appendix D).

The reason I insert the factor $1/N!$ is that the particles are *identical*, so interchanging two particles would appear to add a new way of arranging the N particles. But it does not because the new configuration is indistinguishable from the first. We must sum up distinguishable *states* in the partition function. Indeed, in German, Z stands for Zustandssumme (state sum). It is quantum mechanics that teaches us that particles, including atoms and molecules, are indistinguishable.

Problem 15.1

The Tafel coefficients are not independent. Show that at de-electronation,

$$a = -b \log i_0$$

and at electronation,

$$a = b \log i_0$$

and hence,

$$\eta = \pm b \log \frac{i_0}{i}$$

the anodic overpotential being positive and the cathodic overpotential being negative. This is most often how the Tafel equation is expressed.

Solution

At de-electronation,

$$i = i_0 \exp\left(\frac{\alpha n F}{RT}\eta\right)$$

Taking natural logs and converting to base 10,

$$\log i = \log i_0 + \frac{\alpha n F}{2.3RT}\eta$$

which is

$$\eta = a + b \log i$$

with

$$a = -\frac{2.3RT}{\alpha n F} \log i_0$$

and

$$b = \frac{2.3RT}{\alpha n F} > 0$$

It follows that

$$a = -b \log i_0$$

and

$$\eta = -b\left(\log i_0 - \log i\right) = -b \log \frac{i_0}{i} > 0$$

At electronation,

$$i = i_0 \exp\left(-\frac{(1-\alpha) n F}{RT}\eta\right)$$

Taking natural logs and converting to base 10,

$$\log i = \log i_0 - \frac{(1-\alpha)\,nF}{2.3RT}\eta$$

which is

$$\eta = a - b\log i$$

with

$$a = \frac{2.3RT}{(1-\alpha)\,nF}\log i_0$$

and

$$b = \frac{2.3RT}{(1-\alpha)\,nF} > 0$$

It follows that

$$a = b\log i_0$$

and

$$\eta = b\left(\log i_0 - \log i\right) = b\log\frac{i_0}{i} < 0$$

Problem 15.2

Returning to Problem 1.2, using the voltage you have measured make an estimate of the concentration of zinc ions dissolved in the lemon juice in the steady state. You should assume that the partial pressure of hydrogen gas is one bar. The pH of lemon juice is about 2.

Solution

At the anode, the reaction is

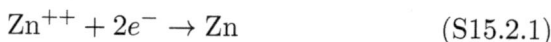

$$Zn^{++} + 2e^- \rightarrow Zn \qquad (S15.2.1)$$

Zn^{++} is the oxidised state and metal Zn is the reduced state. According to the Nernst equation,

$$\mathcal{E}_a = \mathcal{E}_0 + \frac{RT}{2F}\ln h_{Zn^{++}}$$

if we take the activity of the metal to be one, and $h_{Zn^{++}}$ is the quantity we are looking for. You can look up $\mathcal{E}_0$ in the electrochemical series (it's –0.763 V). At the cathode is reduction of the proton,

$$2H^+ + 2e^- \rightarrow H_2$$

and according to Equation (16.2.6),

$$\mathcal{E}_c = -2.3\frac{RT}{F}\left(pH + \frac{1}{2}\log p_{H_2}\right)$$

Suppose that the voltage you measure is

$$V = \mathcal{E}_c - \mathcal{E}_a$$

Then you find that

$$\ln h_{Zn^{++}} = -\frac{2F}{RT}\left(V - 0.76 + 2.3\frac{RT}{F}\left(pH + \frac{1}{2}\log p_{H_2}\right)\right)$$

Let us assume that the partial pressure of hydrogen gas is one bar (how reasonable is that?). Putting in numbers,

$$F = 96485 \text{ C mol}^{-1}, \quad R = 8.314 \text{ J mol}^{-1} \text{ K}^{-1},$$
$$T = 298 \text{ K}, \quad pH = 2$$
$$\log h_{Zn^{++}} = -\frac{1}{2.3}77.9\,(V - 0.76 + 0.59pH) = 9.8$$

since my lemon lamp recorded 0.93 V.

So the zinc concentration is about 10^{-10}. That's essentially zero zinc, so you can eat the lemon afterwards. Note that the greater the voltage, the less zinc is in solution. That seems counterintuitive, but you can see it in the Nernst equation: for a given $\mathcal{E}_c$ since $\log h_{Zn^{++}}$ is *negative*, the more zinc is in solution, the less base is the anode potential. Another way to see it is that a high concentration of Zn^{++} drives the anode reaction (S15.2.1) to the right.

Problem 16.1

A piece of iron is corroding in neutral aerated fresh water. Assuming equal anode and cathode areas, estimate the corrosion potential, the corrosion current density and the uniform corrosion rate in mm per year. Take the activity of the iron ions to be 10^{-6}.

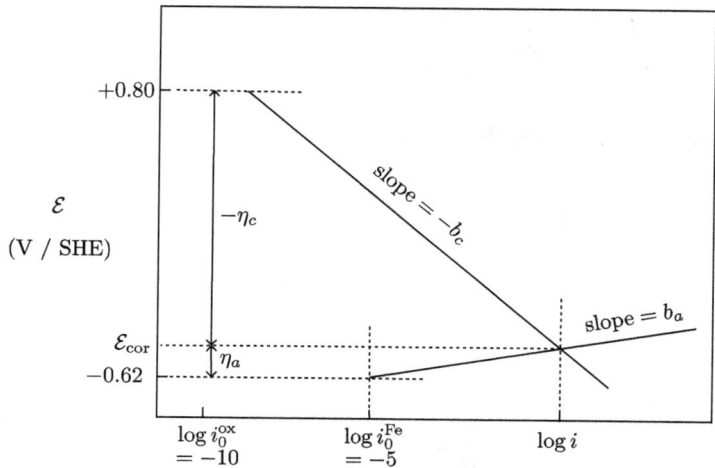

Fig. C.4.   An Evans diagram to illustrate the solution to Problem 16.1.

Solution

First draw the Evans diagram (see Figure C.4). Since the water is neutral and aerated, the cathode reaction will be the reduction of dissolved oxygen.

I will use these data (see Table 14.1).

(1) Anodic Tafel slope, $b_a = 0.05$
(2) Cathodic Tafel slope, $b_c = 0.12$
(3) Cathode exchange current density: $\log i_0^{ox} = -10$
(4) Anode exchange current density: $\log i_0^{Fe} = -5$

The cathodic overpotential is 0.8 V/SHE, Equation (16.2.4). The standard electromotance for reduction of Fe^{++} is –0.44 V. Assuming iron ion activity of 10^{-6}, we find from the Nernst equation

$$\mathcal{E}_a = \mathcal{E}_0 + \frac{RT}{2F} \ln 10^{-6} = -0.62 \text{ V/SHE}$$

Using results from Problem 15.1 the overpotentials at anode and cathode respectively are

$$\eta_a = a_a + b_a \log i = b_a \left(\log i - \log i_0^{Fe} \right) > 0$$
$$\eta_c = a_c + b_c \log i = b_c \left(\log i_0^{ox} - \log i \right) < 0$$

where i is the corrosion current density we are looking for. I also see from the Evans diagram that

$$\eta_a - \eta_c = 0.8 + 0.62 = 1.42 \quad V$$

I can solve these three equations for $\log i$ and I get

$$1.42 = 0.05 \log i - 0.05 \times (-5) + 0.12 \log i - 0.12 \times (-10)$$

giving

$$\log i = -0.18$$

So,

$$i = 0.67 \; Am^{-2}$$

and the estimated uniform corrosion rate is 0.67 mm per year.

We see from the Evans diagram that $\mathcal{E} = -0.62 + \eta_a$ and $\eta_a = 0.05 \times (-1.42 + 5)$. So the corrosion potential is

$$\mathcal{E} = -0.44 \; V$$

Alternatively, you can treat this problem graphically, either by careful plotting or by regarding the two straight lines on the Evans diagram as

$$y_a = y_{0,a} + b_a (x - x_{0,a})$$
$$y_c = y_{0,c} - b_c (x - x_{0,c})$$

where y is the ordinate, $\mathcal{E}$, and x, the abscissa, $\log i$. We have in this instance

$$y_{0,a} = -0.62 \; V, \quad y_{0,c} = 0.8 \; V,$$
$$x_{0,a} = \log i_0^{Fe} = -5, \quad x_{0,c} = \log i_0^{ox} = -10$$

Then you set the two right-hand sides equal to locate the intersection and arrive at

$$\log i = \frac{1}{b_a + b_c} (0.8 + 0.62 + b_a \log i_0^{Fe} + b_c \log i_0^{ox}) = -0.18$$

as before.

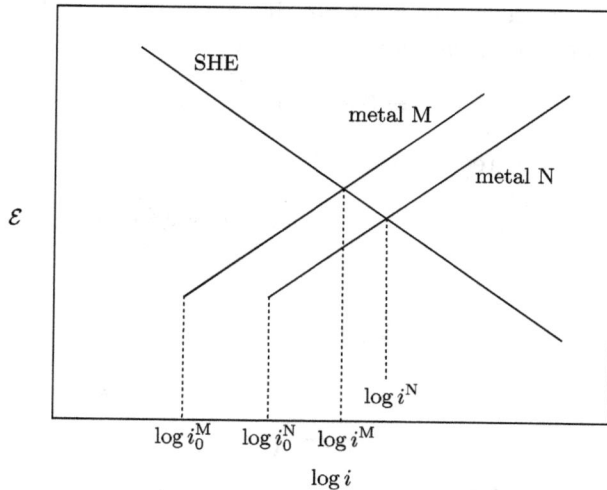

Fig. C.5.   An Evans diagram to illustrate the solution to Problem 16.2.

Problem 16.2

A metal, M, is used as an anode electrode in electrical contact with a standard hydrogen electrode. It is found that the current density is i^M. The electrode is now replaced with a different metal, N, having the same dimensions and the same Tafel slope as the electrode, M. The measured current density is now i^N which is greater than i^M. If the Tafel slope of the metals, M and N, is denoted b and the Tafel slope for the SHE is $-b^{SHE} < 0$, and if the exchange current densities of metals M and N are denoted i_0^M and i_0^N, show that

$$\log \frac{i_0^M}{i_0^N} = \left(1 + \frac{b}{b^{SHE}}\right) \log \frac{i^M}{i^N}$$

Solution

First draw an appropriate Evans diagram (see Figure C.5).

Using the solution to Problem 15.1, the anodic overpotential is

$$\eta = -b\left(\log i_0 - \log i\right)$$

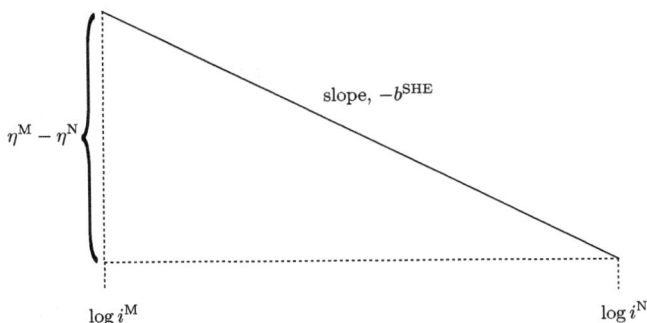

Fig. C.6. A construction to illustrate the solution to Problem 16.2.

Then on metals M and N respectively we have, with $b > 0$,

$$\eta^{M} = -b \left(\log i_0^{M} - \log i^{M} \right) > 0$$
$$\eta^{N} = -b \left(\log i_0^{N} - \log i^{N} \right) > 0$$

Now we observe from Figure C.6 that using the cathodic Tafel slope of the SHE

$$\eta^{M} - \eta^{N} = -b^{SHE} \left(\log i^{M} - \log i^{N} \right)$$

Then,

$$-b \left(\log i_0^{M} - \log i^{M} \right) + b \left(\log i_0^{N} - \log i^{N} \right) = -b^{SHE} \left(\log i^{M} - \log i^{M} \right)$$

which is

$$\log \frac{i^{M}}{i^{N}} \left(b + b^{SHE} \right) = b \log \frac{i_0^{M}}{i_0^{N}}$$

Finally, as required,

$$\log \frac{i_0^{M}}{i_0^{N}} = \left(1 + \frac{b}{b^{SHE}} \right) \log \frac{i^{M}}{i^{N}}$$

Problem 17.1

The slope of line 5 in Figure 17.3 is one-third that of line 4. Given that the same number of electrons are involved, what does this immediately tell you?

Solution

That there are three times as many protons involved in reaction 5 than in reaction 4.

Problem 17.2

Figure C.7 shows a Pourbaix diagram for copper in water at 25°C.
   Find the equation of the boundary that separates Cu from Cu_2O. The standard free enthalpy of Cu_2O is –148 kJ mol^{-1} and that of pure water is –237 kJ mol^{-1}. That's all you need to know.

Solution

The reaction is

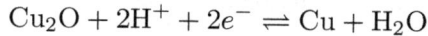

$$Cu_2O + 2H^+ + 2e^- \rightleftharpoons Cu + H_2O$$

The standard free enthalpy of the reaction at 298 K is

$$\Delta G^\circ = -237 - (-148) = -89 \text{ kJ mol}^{-1}$$

and so the standard electromotance is $\mathcal{E}^0 = -\Delta G^\circ/nF = 89000/(2 \times 96485) = 0.45$ V/SHE. Then using the Nernst equation,

$$\mathcal{E} = \mathcal{E}^0 + \frac{RT}{2F} \ln h_{H+}^2 = \mathcal{E}^0 + \frac{2.3RT}{F} \log h_{H+} = 0.45 - 0.059pH \quad \text{V/SHE}$$

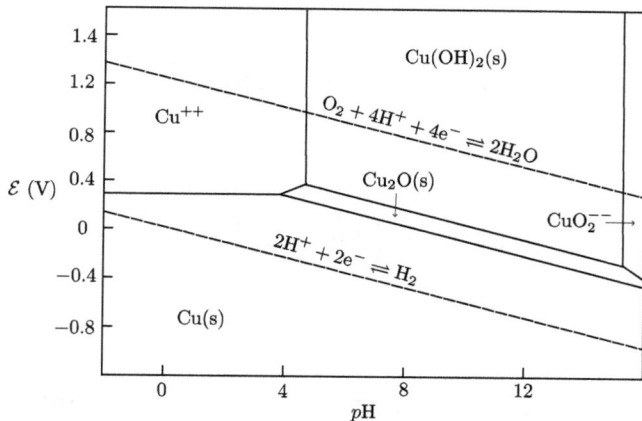

Fig. C.7.   Pourbaix diagram for Cu in water.
Source: Metallos, CC BY-SA 4.0, via Wikimedia Commons.

Appendix D

Further Reading

J. M. West, *Basic Corrosion and Oxidation*, 2nd edn. (Chichester, Ellis Horwood, 1986).

- This is the book from which I learned corrosion as an undergraduate. Indeed John West was our lecturer. I know of no better or more correct and rigorous text, but sadly it is out of print.

J. O'M. Bockris and A. K. N. Reddy, *Modern Electrochemistry* (Plenum, New York, 1970).

- This long and rambling text is a pedagogical masterpiece, still in print after more than 50 years.

W. R. Fawcett, *Liquids, Solutions, and Interfaces* (OUP, Oxford, 2004).

- An excellent and authoritative text very much aimed at the physical chemist.

W. Schmickler and E. Santos, *Interfacial Electrochemistry*, 2nd edn. (Springer, Berlin, 2010).

- This book is the closest to my own text, given that Wolfgang Schmickler is originally a theoretical physicist. However, the selection of topics is sufficiently different for me to justify the book that you are reading.

W. J. Moore, *Physical Chemistry*, 5th edn. (Longman, London, 1972).

- I had this text as an undergraduate and to my mind it is superb. There is a very clear and rigorous introduction to statistical mechanics.

N. D. Lang and W. Kohn, "Theory of metal surfaces," *Physical Review B*, **1**, 4555 (1970); *ibid.*, **3**, 1215 (1971).

- These are the classic two papers on "jellium" and the density functional theory of surfaces. Kohn invented DFT and Lang wrote one of the most accessible reviews of it.

J. Cheng and M. Sprik, "Alignment of electronic energy levels at electrochemical interfaces," *Physical Chemistry Chemical Physics*, **14**, 11245 (2012).

- This was the starting point of some of my puzzlement in the subject. A very precise statement of the variety of "potential" that I introduce here as well as vital treatment of the relation of those which are measurable to what can be calculated using first principles quantum mechanics.

J. W. Gibbs, *Collected Works*, Vol. 1 (New Haven Connecticut, Yale University Press, 1957), pp. 219–331.

- I include this for your retirement. It is far from an easy read and Gibbs really requires a modern day interpreter. All his notation is unfamiliar to the present reader and like the book that Alice's sister was reading, there are no pictures.

J. W. Cahn, Thermodynamics of solid and fluid surfaces, in W. C. Johnson and J. M. Blakely (Eds.) *Interfacial Segregation* (Metals Park Ohio, American Society for Metals, 1979), p. 3.

- So Cahn has stepped in, in this one instance, to give us both an interpretation and even an improvement on Gibbs's interfacial thermodynamics.

M. Finnis, *Thermodynamics of Interfaces* (Unpublished Lecture Notes, Imperial College London and Max-Planck-Institut für Eisenforschung, Düsseldorf).

- I have drawn on these notes (with the author's permission) to write the first section of my Chapter 11.

H.-J. Butt, K. Graf and M. Kappl, *Physics and Chemistry of Interfaces*, 2nd edn. (Weinheim, WILEY-VCH Verlag GmbH, 2006), Chapter 3.

- This is a very useful text for undergraduates and graduates, closer than Fawcett to the physicist. They use Gibbs's dividing surface rather than Cahn's and Finnis's theory.

K. Denbigh, *The Principles of Chemical Equilibrium*, 4th edn. (Cambridge, Cambridge University Press, 1981).

- This was Dr Roy Buckle's recommended text when I was a student and it is superb — total rigour. If we had to eliminate all but one text in thermodynamics, this is the one that I should hope would survive. Here is all you will need in classical and chemical thermodynamics and a detailed treatment of activity in non-ideal solutions and the conventions used to make calculations. He has very clear and very extensive chapters on statistical mechanics.

A. Sommerfeld, *Thermodynamics and Statistical Mechanics*. Lectures on Theoretical Physics, Vol. 5. Translated by J. Kestin (Academic Press, New York, 1956).

- Of course, this book would also have to survive the cull, given that it belongs to his extraordinary series in theoretical physics. Students are bound to be confused by the two approaches to statistical mechanics: Gibbs's and Boltzmann's; here is where you will find them unravelled. Sommerfeld must have been the master of pedagogy. Not only did he teach in Munich for many decades, he also supervised some of the most famous names in quantum mechanics. He was also the most skilled mathematician and delights in finding the clearest and most logical paths to where he wishes to take us. Here is a typical quote: "The path taken by Einstein in 1905 in the discovery of the special theory of relativity was steep and difficult... The path that we shall take is wide and effortless." I have all six volumes and I have used all of them in my own teaching.

Index

A

absolute single electrode potential, 71
acidity, 166
activated complex, 113, 121, 129
activity, 28, 203, 206, 255
  Henrian, 166
      practical, 56
  ion
      mean, 68, 70, 218
      thermodynamic, 217
      working, 218
  kink, 137, 144
  Raoultian, 209
activity coefficient, 120, 203–204, 206,
  209–210, 213
  ion, 216
      mean, 57, 70, 216
  practical, 213, 217
  rational, 213
air gap, 67, 74
alloy, 57, 82, 189, 194, 200, 206–207
aluminium, 179
anion, 98, 101
anode, 1, 60, 63, 108, 163, 173, 175,
  255, 258
anti-muon, 41
antiphase boundary, 83
apology, 28
Arrhenius, Svante, 122, 128
atom fraction, 203, 211

attempt frequency, 114–115, 130
  de-electronation, 136–137, 143–144,
    146
  electronation, 135–136, 143–144,
    146
  in statistical mechanics, 128,
    130
auxiliary electrode, *see* counter
  electrode, 148
Avogadro constant, 28–29, 119, 195,
  250

B

base metal, 64, 152
battery, 61
Beltramo, Guillermo, 98, 100
Bilby, Bruce, 161
Bockris point, 35, 39, 44, 46, 48, 71,
  145
Bockris, John, viii, 1, 5, 17–18, 32–33,
  41, 78, 92, 101, 105, 111, 113, 124,
  261
Boltzmann constant, 109, 119, 195,
  232
Boltzmann, Ludwig, 207, 231
Brearley, Harry, 179
Buckle, E. Roy, 263
Butler–Volmer equation, 108, 112,
  155, 157–158, 160

C

Cahn, John, 77–78, 81, 83, 87, 89, 262
calomel, 71, 219, 230
   electrode, 219
      saturated, 230
      standard, 98, 231
capacitance, 19, 98, 104, 231
capacitor, 9, 16, 19, 77, 92, 101, 159, 231
capacity, 92, 104
   Gouy–Chapman, 105
   Helmholtz, 105
capillarity, 91
cathode, 3, 60, 163, 173, 175, 183, 255
cation, 44, 60, 98, 101, 107
cell diagram, 5, 57, 68, 74, 230
Chapman, David, 32, 101
charge number, 28, 97, 103, 214
chemical potential, 12, 23, 27, 45, 49, 78, 82–83, 195, 201, 215, 245, 248
   electrolyte, 217
   electron, 14–15, 22, 25, 234, 236
   hydrogen gas, 74
   ideal mixture, 208
   solute, 210
   standard, 27, 45–46, 165, 213
      electron, 56, 68–70, 72–74
      ion, table of, 63
      unimolal, 57, 213
Cheng, Jun, 48–49, 75, 262
chloride, 108, 248
   evil, 188
   ion, 108
   oxidation, 149
chromium, 179
chronocoulometric method, 92
clash
   of notation, 28
Clausius, Rudolf, 192
cobalt, 109
coexistence line, 83
cohesive energy
   of metal, 146
coinage metals, 63
compensation potential, 19, 67–69, 73

complexions, 231
conjugate variables, 191
contact potential, 18, 23–24
contact potential difference, 19, 238
copper, 1, 57, 72, 147, 173, 206, 230
   Pourbaix diagram, 260
corrosion, 1, 55, 107, 110
   free enthalpy change, 174
   immune, 177
   potential, *see* potential, 108
   protection, 179
   rate, *see* rate, 257
   uniform, 108, 174, 187
counter electrode, 148
coup de grâce, 129
crevice attack, 176, 187
current density, 108, 149, 158, 168, 175, 258
   corrosion, 175, 255
   de-electronation, 156, 160
      equilibrium, 138, 143
   electronation, 156, 160
      equilibrium, 136, 143
   exchange, 143, 176
      de-electronation, 256
      electronation, 256
      equilibrium, 146, 149, 168
      table of, 150
   oxidation, 143
   reduction, 143

D

Dalton's Law, 207
Davies, Graeme, 188
de-electronation, 1, 60, 136, 138, 146, 160, 174–175, 252
Debye length, 104–105
degrees of freedom, 82, 122, 250
Denbigh, Kenneth, 62, 201, 263
density functional theory, 12, 20, 71, 234
detailed balancing, 133
Devine, John, 188
differential aeration, 176
differential capacity, 95
differential resistivity, 158

diffusion, 169, 189, 196
dipole moment, 16
dipole potential, 22, 29, 31
  water, 71, 74, 244
dipole potential difference, 33, 69, 74
dividing surface, 112, 124, 127, 129
  Gibbs's, 90
double layer, 11, 16, 46, 98, 135, 172, 180
  Gouy–Chapman, 32
  Helmholtz, 101
  Stern, 32

E

Ehrenfest, Paul, 231
Ehrenfest-Afanassjewa, Tatjana, 231
Einstein
  frequency, 130
  summation, 78
Einstein, Albert, 263
electric field, 16, 32, 35–36, 135, 151, 238
electric potential, 12
electrocapillary equation, 92–93
electrochemical potential, 24, 27–28, 46, 60, 67, 72, 96, 249
  electron, 23–24, 56, 59, 68, 71–72, 96
    in metal, 133
  ion, 45, 48, 133
electrochemical series, 63, 71, 134, 155, 173, 181, 248
  table, 63
electrode potential, *see* single electrode potential
electrolyte, 5
electrolyte chemical potential, *see* chemical potential
electromotance, *see* electromotive force
electromotive force, 60, 175, 177
  driving, 174
  hydrogen evolution reaction, 167
  reduction of dissolved oxygen, 167
  standard, 63, 65, 220
    reduction of ferric ion, 256

electromotive force series, *see* electrochemical series
electron density, 13
electron volt, 12
electronation, 3, 135, 138, 146, 160, 174–175, 252
electronegativity, 236
enthalpy, 190
  activation
    critical, 113
    standard, 123
  partial molar, 198, 204
entropy, 81, 83, 190, 207, 231, 250
  activation, 129
    standard, 123
  Boltzmann formula, 207
  partial molar, 198
Entwisle, Tony, 108, 188
equilibrium, 16, 23, 59, 82, 121, 133, 192, 200
equilibrium constant, 54, 120
equipartition, 250
error, viii, 77, 81–82, 90, 213
ethanol, 244
Euler's theorem, 198
Euler–Lagrange equation, 14, 236
Evans diagram, 161, 163, 168, 174, 176–177, 179, 256–258
Evans, Ulick, 176
excess, 78, 84, 86, 93, 97, 245, 248
  interfacial, 90
  interfacial charge, 94, 96, 248
  interfacial free enthalpy, 80
exchange, 41
exchange and correlation, 14, 235
exchange current density, *see* current density
external potential, 13–14, 234–235
extrathermodynamic, 70–71
Eyring, Henry, 109

F

Faraday constant, 28, 47, 94, 136
Faraday, Michael, 1, 3, 24

Fawcett, W. Ronald, viii, 49, 65, 68, 145, 152, 261
Fermi energy, 15, 20, 23, 56, 73–74, 94, 96, 237
Fermi level, *see* Fermi energy
Finnis, Mike, 77–78, 80, 83, 87, 91, 262
free energy, 190
  activation, 129
  partial molar, 198
  statistical mechanics, 119
free enthalpy, 27, 78, 190
  activation, 113–114, 140–141
    chemical, 135, 141, 143
    de-electronation, 136
    electronation, 135
    standard, 123
  mixing, 208
  oxidation, 133
  partial molar, 198
  standard
    cuprous oxide, 260
    formation, 55
    reaction, 54, 59, 220
    water, 260
free enthalpy of solvation, 28, 65, 241
  cation, 146
  proton, 70, 244
  real, 43
    ion, 50
  standard, 28, 49
    ion, 49
frequency prefactor, *see* attempt freqency
fuel cell, 61, 157
functional derivative, 14, 232–233, 235
fundamental theorem of calculus, 233

G

gallium, 247–248
gallium chloride, 247
Galvani potential, 22, 31, 46, 91–92, 105, 248

Galvani potential difference, 17, 74, 97, 247
galvanisation, 177
Gibbs adsorption equation, 81, 86, 93
Gibbs adsorption isotherm, 82, 84, 90, 96
Gibbs free energy, 190
Gibbs phase rule, 82, 199
Gibbs's paradox, 207
Gibbs, Josiah Willard, 77, 82, 90, 201, 262
  ghost, 68
Gibbs–Duhem equation, 82–86, 199, 206
Gibbs–Helmholtz equation, 204
Gouy, Louis, 32, 101
Grahame, David, 95
grain boundary, 83
graphite, 148
ground state, 13–14, 121, 234–235

H

half cell, 56, 65, 231, 248
  single potential, 65
half life, 110
harmonic approximation, 128
Hartree energy, 14, 232–233
Hartree potential, 14, 232
Heine, Volker, 84
Hellmann–Feynman theorem, 234
Helmholtz free energy, 190
Helmholtz, Herman von, 11
Henrian standard state, *see* standard state
Henry's law, 204, 210
HER, *see* hydrogen evolution reaction
Heyrovsky reaction, 170
hydrochloric acid, 68
hydrogen economy, 168, 178
hydrogen electrode, 57
  standard, 8, 61–62, 68, 72, 94, 147, 155, 167, 220, 230, 242, 247, 258
hydrogen embrittlement, 177

hydrogen evolution reaction, 167–168, 173–174, 176–177
hydrogen ion, *see* proton
hydronium ion, 169–171

I

ideal gas, 71, 119, 206, 208
ideal mixture, 206
ideal solution, 206
image force, 20, 33
image potential, 32, 39, 41
immunity, 182
impedance spectroscopy, 98
inner potential difference, 12, 15, 19, 22, 25, 29, 94, 96
   metal-electrolyte, 56
interfacial tension, 78, 91–92
   gallium, 248
   mercury, 95
   water, 245
internal energy, 81, 189, 251
interphase, 11, 31, 77, 85, 92, 95–96, 101
   capacitance, 101
   differential resistivity, 158
   electric field, 151
   ohmic, 158
   rectifying, 159
   thermodynamics, 77
ionisation energy, *see* ionisation potential
ionisation potential, 43, 65, 70, 146, 241–242
   water, 242
IR drop, 176
iron, 43, 56, 176, 181, 255
   Pourbaix diagram, 183

J

Jacob's ladder, 134
Janak, James, 23

K

Kelvin probe, 19–20, 29, 31, 67, 238
Kelvin, Lord, 19, 24

Kenrick's apparatus, 29, 67, 69, 73
Kittel, Charles, 20
Kohn, Walter, 12, 21, 262

L

ladder diagram, 6, 59
Lagrange multiplier, 14, 231
Lang, Norton, 12, 21, 262
layer quantities, 81
LCR circuit, 77
lead, 176, 206
Legendre transformation, 191
lemon lamp, 1, 5–6, 63, 108, 178, 229, 231, 255
Lippmann equation, 91–92, 96–97, 248
Lippmann, Gabriel, 91, 98
Lovins, Amory, 56
Luggin capillary, 148

M

magnesium, 64, 179
Maxwell, James Clerk, 77
mean inner potential, 15, 71
mean ion activity, *see* activity
mean ion activity coefficient, *see* activity coefficient
mercury, 67–68, 91–92, 219, 230
mixed potential, *see* corrosion potential
molality, 212
   mean ion, 216
molar mass, 203, 211
   relative, 203
mole, 29, 195, 211
mole fraction, 203–204, 211
molecular dynamics, 110, 131, 171
molecular weight, 203, 211
Moore, Walter, 123, 134, 169

N

Nernst equation, 65, 147, 155, 157, 164–166, 172, 174, 206, 248, 254, 256, 260
nickel, 109, 206

noble metals, 63
non polarisable electrode, *see*
    reversible single electrode
normal mode, 127

O

OHP, *see* outer Helmholtz plane
open circuit voltage, 57
outer Helmholtz plane, 135–136, 140,
    142, 145
outer potential, 22, 31, 39
overpotential, 138, 140–141, 151, 155,
    168, 172, 256, 258
    high limit, 159
    hydrogen evolution reaction, 168
    low limit, 158
oxidation, 1, 55, 58–59, 138, 229
oxide, 55

P

Parsons, Roger, 46
partition function, 116, 120, 124,
    128–129, 251
    canonical, 118
    microcanonical, 117
        electronic, 118
        rotational, 118
        vibrational, 118, 122
    standard state, 120
passivation, 179
passivation current density, 181
patch field, 19
peak passivation potential, 180
permittivity, 103
perpetual motion machine, 237
pH, 166, 168, 177, 181
phenolphthalein, 163
photoelectric effect, 20
photon, 109
pitting, 108, 187
Planck constant, 109, 114
platinum, 57, 60, 165, 230, 247
Poisson equation, 18, 103
Poisson–Boltzmann equation, 103
polarisable, 9

polarisation curve, 175–176, 179
positron, 41
potassium chloride, 230
potassium hexacyanoferrate, 163
potential, 12
    corrosion, 108, 163, 171, 173–175,
        178, 257
potential difference, 6, 159
    equilibrium, 57, 135, 147, 152, 155,
        172
        standard, 147
    interphase, 138
potential energy, 12, 112
    activation, 129
potential of zero charge, 92, 95, 105,
    151–152
Pourbaix diagram, 179, 181–182
    copper, 260
    iron, 183
proton, 3, 8, 28, 41, 60, 68, 94, 102,
    166, 170–171, 260
    elementary charge, 1, 103
    real potential, *see* real potential
    reduction, 159, 163, 255
    work function, *see* work function
PZC, *see* potential of zero charge

Q

quantum mechanics, 12, 109, 206
    tunnelling, 129, 180
quarks, 71

R

Raoult's law, 204, 209
Raoultian standard state, *see*
    standard state
rate
    corrosion, 181
        uniform, 257
    de-electronation, 136
        equilibrium, 136
    electronation, 136
        equilibrium, 136
rate coefficient, 109, 115, 122, 129,
    134

de-electronation, 146, 180
electronation, 146
rate equation, 110
reaction
order of, 109
reaction coordinate, 112, 137
real potential, 27, 29, 43, 49, 67, 70
proton, 67, 75
standard, 44
Reddy, Amulya, viii, 1, 5, 18, 32–33,
41, 92, 101, 105, 111, 261
redox
couple, 44, 63, 150, 242
standard electromotive force,
242
equilibrium, 164
reaction, 61, 142, 147, 165
reduction, 3, 59, 138, 229
reduction of dissolved oxygen,
173–175, 177, 256
reservoir, 27–28, 46, 71, 192, 196,
201–202
resistor, 9, 77, 159, 231
reversible single electrode, 9, 159
Rugg, Dave, 188

S

Sackur–Tetrode formula, 252
saddle point, 112, 114–115, 122
Santos, Elizabeth, viii, 43, 92, 98,
100, 105, 144, 158, 161, 242, 261
scanning tunnelling microscopy, 100
Schmickler, Wolfgang, viii, 43, 92, 98,
105, 144, 158, 161, 242, 261
segregation, 86
SHE, *see* standard hydrogen electrode
silver, 57, 72, 165, 173–175
silver chloride, 57
single electrode potential, 7, 9, 61–62,
71, 177, 181
absolute, *see* absolute single
electrode potential
equilibrium, 175, 182
standard, 63, 151, 155, 165
soap, 84

solute, 28, 49, 203–204, 206, 210–211,
213
solvation shell, 28
solvent, 28, 46, 49, 56, 83, 85–86,
96–97, 203–206, 210–215
Sommerfeld, Arnold, 252, 263
specific adsorption, 98
sphere
inner, 101
outer, 101
spinodal decomposition, 206
Sprik, Michiel, 48–49, 75, 262
stacking fault, 83
stainless metal, 152, 179
stainless steel, 179, 181
standard hydrogen electrode, *see*
hydrogen electrode
standard state, 23, 55, 203–204
choosing, 209
Henrian, 46, 210
one bar, 209
pure solvent, 209
pure substance, 207, 209
Raoultian, 46, 210–211
unimolal, 56, 210, 213
unspecified, 208
Stern, Otto, 11, 32
Stirling approximation, 119, 207, 251
sulphuric acid, 177, 181
surface, 11
dipole, *see* dipole potential
surface film, 179
surface tension, *see* interfacial tension
symmetry factor, 112, 135, 138, 141,
144, 146–147, 158
system, 78, 189
layer, 85

T

Tafel law, 160, 168, 175, 253
Tafel line, 176
Tafel reaction, 170
Tafel slope, 168, 175–176, 180–181,
252, 256, 258
standard hydrogen electrode, 259

Tafel, Julius, 168
thallium, 206
thermal de Broglie wavelength, 118,
    249–250
thermionic emission, 20
tin, 206
titanium, 179
total energy, 13, 21, 71, 130,
    234–235
transfer coefficient, 138
transition state theory, 109, 114
transmission coefficient, 131, 138
Trasatti, Sergio, 71, 73, 152
triple point, 82
tungsten, 20, 237
twin, 83

U

unimolal standard state, *see* standard
    state

V

vacuum level, 20–21, 51, 71
van 't Hoff isotherm, 54, 123
Vineyard, George, 123
Voigt, Woldemar, 77
Volmer reaction, 169
Volta potential difference, 22, 31, 40,
    67, 69, 73
    metal-electrolyte, 146
voltammetry, 149
    cyclic, 98

voltammogram, 98
volume
    partial molar, 198

W

water, 75, 82, 84, 98, 167, 173, 175,
    177, 241, 244, 255
    decomposition, 177
    dipole potential, *see* dipole
        potential
    interfacial tension, *see* interfacial
        tension
    standard free enthalpy, *see* free
        enthalpy
Wert, Charles, 123
West, John, 17, 73, 107, 137, 163,
    181, 188, 261
work function, 20, 24–25, 29, 33, 44,
    70, 237–238, 242
    electrolyte, 43, 241
    ion, 48
    proton in water, 68, 70, 244
    standard, 44

Z

Zener, Clarence, 123
zero of energy, 12, 27, 29, 45, 55, 71,
    117, 120
zinc, 1, 6, 176, 230, 254
zinc sulphate, 8
Zundel ion, 171
Zustandssumme, 252

www.ingramcontent.com/pod-product-compliance
Lightning Source LLC
Chambersburg PA
CBHW061629220326
41598CB00026BA/3940